1998

Applied Population Ecology

Applied Population Ecology
A Supply–Demand Approach

Andrew Paul Gutierrez
Environmental Science Policy and Management
University of California, Berkeley, CA

with the assistance of

C. K. Ellis

John Wiley & Sons, Inc.
NEW YORK / CHICHESTER / BRISBANE / TORONTO / SINGAPORE

Library of Congress Cataloging in Publication Data:

Gutierrez, A. P.
    Applied population ecology: a supply-demand approach / Andrew
Paul Gutierrez with the assistance of C. K. Ellis.
        p.   cm.
    Includes bibliographical references (p.   ).
    ISBN 0-471-13586-0 (alk. paper)
    1. Population biology—Mathematical models.   2. Ecology—
Mathematical models.   3. Insect populations—Mathematical models.
4. Agricultural ecology—Mathematical models.   I. Title.
    QH352.G87   1986
    574.5′248′011—dc20   95-43654

Printed in the United States of America

10 9 8 7 6 5 4 3 2 1

This Book is Dedicated to My Professors:
Robert van den Bosch *(1925–1978)*
Carl B. Huffaker *(1914–1995)*
To
Neil E. Gilbert,
and of Course, My Friend and Wife
Marina

On Nature
natura non facit saltum
—Alexander Pope, in *An Essay on Man*

On the Species *Homo sapiens*

Our problem is that we are too smart for our own good, and for that matter, the good of the biosphere. The basic problem is that our brains enable us to evaluate, plan, and execute. Thus, while all other creatures are programmed by nature and subject to her whims, we have our gray computer to motivate, for good or evil, our chemical engine. Indeed, matters have progressed to the point where we attempt to operate independently of nature, challenging her dominance of the biosphere. This is a game we simply cannot win, and in trying we have set in train a series of events that has brought increasing chaos to the planet.

—Robert van den Bosch, 1978

Acknowledgments

This book has been in progress for several years and could not have been completed without the generous assistance of many people. First and foremost is Neil Gilbert who got the ball rolling in the right direction. In historical perspective, Y. H. Wang (co-authored Appendices II and IV) and J. U. Baumgärtner were significant contributors to the ideas presented herein. C. K. Ellis labored long on our joint work. He prepared most of the illustrations and made many significant technical contributions to model development. This contribution is acknowledged on the book cover. Uri Regev is both thanked and blamed for the economic influences in the book. Maurizio Severini wrote Appendix I on basic mathematics.

My colleagues Guy Curry, Nick Mills, and Gosta Nachman were thoughtful reviewers. Sebastian Schreiber made significant corrections to the mathematics, but I take responsibility for all remaining errors. Bill Voigt prepared some of the illustrations. Of course, my many doctoral and postdoctoral students endured and stimulated new thinking and were the reason the text was written in the first place.

Few scientists think of agriculture as the chief, or the model science. Yet it was the first science—the mother of all sciences; it remains the science which makes human life possible; . . .
—*André Mayer and Jean Mayer, 1974*

Preface

This book summarizes lecture notes developed for a class in insect ecology I have taught since 1976 at the University of California at Berkeley. It is a how-to book that makes no attempt to review all of the very different approaches to modeling of populations except as they apply to the theme of the effects of resource supply–demand developed herein. Much of what I say has its roots in the ideas of others, some of whom my ignorance of the full literature no doubt offends. I try to provide *entrée* for students of applied ecology whose *forte* is not mathematics to the increasingly difficult literature in population ecology. My goal is to stimulate field ecologists to expand their use of basic mathematics to examine field problems, to enable them to evaluate the literature with a more discerning eye, and to help them solve more efficiently real world problems such as those passionately outlined by van den Bosch (1978) in his book *The Pesticide Conspiracy*.

The models presented here have solid roots in classical population theory but reflect the imprinting of the California school of biological control, of H. S. Smith, R. F. Smith, and my professors to whom this book is dedicated; of C. S. Holling, R. D. Hughes, and N. E. Gilbert; and of R. M. May, M. P. Hassell, and W. W. Murdoch who have led the field in theoretical directions and provided useful opposition. Of special mention is the work of C. T. DeWit and colleagues in The Netherlands whose applied work influenced me by rediscovery, and, of course, A. A. Berryman who always had interesting controversy at hand.

The first chapter outlines some major arguments that are considered part of the common ecological wisdom and sets the tone for the remaining chapters. The second chapter examines aspects of sampling theory that have had positive impact in applied ecology and owes much to the work of my former student

(L. T. Wilson). Chapter 3 reviews the theory and practice of poikilotherm development that is a keystone for realistic modeling of plant–herbivore–natural enemy interactions in nature. Curry and Feldman (1987) have a first rate review of this theory. Chapter 4 reviews the basis of life table theory and practice, and introduces the background for the development of time-varying life tables proposed by CSIRO colleague R. D. Hughes. Chapter 5 examines the underpinnings of the functional response (i.e., resource acquisition models) in both traditional and nontraditional ways pointing out the need for an appropriate currency (i.e., mass or energy) to link trophic levels. This is also important for examining the physiological basis of predator search behavior. Chapter 6 shows how the selection of an appropriate currency and functional response model can be used to develop per capita models for growth, development, and reproduction that are general across trophic levels. Chapter 7 introduces the concept and realism of energy driven supply–demand theory in the development of simple single-species population dynamics models. The analytical properties of these simple physiologically based models are reviewed. Chapter 8 extends this development to multitrophic models. Single-species mass- and age-structured models are introduced in Chapter 9 and Nicholson's classic blowfly data are analyzed as an example. Chapters 6–9 are key to development of multitrophic models covered in Chapter 10. Chapter 11 extrapolates the core ideas of the book to the analysis of regional problems, and introduces applications of remote sensing and geographic information technology. Chapter 12 is a polemic on ecosystem sustainability that may be safely ignored.

Care has been taken to include intermediate mathematical steps in derivations of models in all chapters, and Maurizio Severini kindly summarized the relevant mathematics in Appendix I. Appendix II applies these mathematics to many of the population models used in population ecology. It was written in conjunction with former research assistant Y. H. Wang. For those not versed in the arguments of model stability (Appendix III), I have built upon a highly accessible expose by Roughgarden (1979). Appendix IV was revised from Gutierrez and Wang (1984) and reviews some optimization approaches that have been used to analyze complex applied problems in pest management. Hopefully, this book helps bridge the gap between the biological complexity faced by field ecologists and parts of ecological theory useful in resource management.

Contents

Applied Population Ecology

1

Introduction

From a letter of recommendation: ...

The basic problem with population biology is its innate complexity. The subject is
infested with oversimplifications. Alternatively, people who face up to the complexity
either get lost in it, or at best manage to sort out one small corner of the subject.
Consequently, valid non-trivial generalizations are elusive.
—N. E. Gilbert, 1988

This book is about modeling ecosystems from a multitrophic perspective and
applying the understanding gained to the management of resources. The basis
for the approach is that *all organisms are faced with the same problems of*
resource acquisition and allocation, and the functions describing the various
components of these processes have similar shapes across all trophic levels
including the economic one. The currency of these interactions is energy or
biomass, but it is recognized that demands for other essential resources may
also affect the dynamics. A unified model for resource acquisition, allocation,
and population dynamics across all trophic levels is developed (cf., Gutierrez
and Wang 1976). In brief, the same model applies to all organisms despite the
fact that the biology, the parameters, and the units vary considerably. The
model describes underlying physiological and ecological relationships making
it independent of the field data it reproduces (Gilbert et al. 1976; Gutierrez and
Wang 1976). The model links physiology and population dynamics by making
birth–death, growth, and net immigration rates functions of the ratio of supply
of resource acquired to the genetic or maximal demand.

A PERSPECTIVE

I share McIntosh's (1987) desire for true plurality in ecology, and this book
was written to express another point of view on how to examine complicated

1

systems. The book *Ecological Relationships* by Gilbert et al. (1976) was brash in its approach to field population ecology, and suffered the consequences of being ahead of its time (Lawton 1977). For a while, it stood as the major, *albeit* weak, opposition to the mill of purely theoretical papers on population ecology (Wang and Gutierrez 1980; Gilbert 1984). Recently, I discovered the book edited by Casti and Karlqvist (1989), which helped sort out some of the underlying conflict between theoretical and applied ecology in the following perspective.

Field biologists observe nature. Why is a relationship as it is? is a recurring question. The role of theoretical science is to answer the *because* to this question. Aristotle was, among many things, a biologist, and his *theory of causes* (i.e., *material, efficient, formal, and final causes*) sought to explain the real world. Aristotle's theory was replaced by Newton's paradigm of *particle, force, and context* because it promised better prediction and was syntactically more precise—it was a mathematical model. This occurred despite the fact that Newton's theory lacked Aristotle's final cause (*the why*), which is at the heart of biology, *Why does an organism behave as it does and not as something else?* Life is not simply the interaction of particles driven by some force in some contextual framework—it is not simply physics and chemistry. Hence, as Casti and Karlqvist point out, Newton was fortunate that he chose to study the movement of celestial bodies and not the workings of a biological system. Yet despite this deficiency, Newton's paradigm became the *bellwether* of science (Casti and Karlqvist 1989), and sadly, many ecologists for a time forsook the study of nature for the allure of simple untestable *mass action models* about population dynamics (cf. Gilbert et al. 1976).

In his farewell to population ecology, the theorist George Oster (1981) wrote: . . . *Remarkably for a long time most ecologists took these equations* (i.e., simple models of complex biology—my emphasis) *quite seriously as if there were hidden in them some great, but subtle truth about nature. What was lost in the proliferation of papers was that the subtlety was mostly mathematical, and the truth they contained mostly allegorical. Indeed, one of the most pernicious consequences of this flood of theorizing-sans-data was that it lent an aura of respectability to allegorical models.*

Only recently have theoretical ecologists started to ask the *why* of the biology their models sought to describe.

Oster also cautioned biologists that . . . *theory can protect empiricists from fortuitous numerology.* I shall attempt to heed this caution, but admonish that sorting out good theory in ecology is no simple matter as an acceptable unifying theory in ecology is just starting to emerge (e.g., Arditi and Ginzburg 1989; Berryman et al. 1995; DeAngelis et al. 1975; Getz 1991; Gutierrez and Baumgärtner 1984a, b; Gutierrez et al. 1994; Huffaker and Rabb 1984; May 1981; Murdoch 1994; and others). To develop a unifying theory in ecology we must, as Gilbert et al. (1976) point out, unravel ecological patterns and try to generalize them across species, trophic levels, and food webs. They recommended studying a diverse set of case histories and heavily documenting them with field data. Their goal was to draw out the commonalties among the systems and generalize them—to develop from first principles a unifying model, which when parameterized would be independent of the field data but still reproduce them. The models should be independent of the constraints of time and place (Akçakaya et al. 1988). The approach developed here is based on these notions, as well as the core idea that all organisms (*except humans*) have a genetic potential demand used solely for growth, development, and reproduction, and it is the biotic and abiotic factors impinging on them that keep them from reaching this potential (Hughes 1963). The demands of humans for resources are more complex. The consequences of these demands are discussed in Chapter 12. If I am reluctant to use the word theory for my approach, it is likely due to the reverence I have for formal theory.

MODELS IN ECOLOGY

As a scientific discipline develops, it progresses from a descriptive science to a quantitative one, and the arguments are increasingly formulated in the language of mathematics. I, like many of my colleagues in ecology, have been making this transition. Though I use simple mathematics to characterize population interactions, there is no pretense that the exercise is rigorous by mathematical standards.

Most ecologists recognize that no model in biology can capture the full richness of predator–prey interactions, for if it did it would be as difficult to understand as nature itself. The theoretical literature on modeling plant–herbivore and prey–predator interactions may be divided into two distinct categories. The first traces its origins back to the differential equation models of Lotka (1925) and Volterra (1926, 1931) and has largely focused on predation in vertebrates. The second traces its origins back to the difference equation models of Nicholson and Bailey (1935) and accounts for almost all analyses of predation in arthropods (Getz and Gutierrez 1982). The book by Kingsland (1985) is a very readable account of historical developments in theoretical ecology.

Ågrens and Bosatta (1990) state that . . . *ecology is the science of the house*, and separate modeling approaches into engineering and scientific approaches (the *E approach* vs. *S approach*). The engineering-approach they conclude reached its height during the IBP (International Biosphere Program) era in the United States. They suggest that we should now shift back to the *S approach*. I agree that much of modeling as well as strict empiricism has lead to the commonly held perception of nature as a series of specific cases. Some simply believe that ecosystems are chaotic (see Appendix II).

APPLIED ECOLOGY

Applied ecologists are asked to find real world solutions, and models are but one of the tools they use. Models help us formulate our notions (*rightly or wrongly*) about the dynamics of the different species that an ecosystem comprises. These models are most useful when they help us to formulate and to test theory that improve our understanding of trophic interactions, and to manage ecosystems in an environmentally friendly manner. These have been long-term goals of ecologists, but rarely have they been achieved. Progress has by and large been modest, and with a few exceptions has been limited to simplified representations of two species interactions with the plant level mostly ignored. Applied ecologists have tended to become lost in the minutia of the biology, satisfying themselves with empirical studies in natural history, and hiding behind the assumption that nature is too complicated to model. Those who have ventured into quantitative areas have stressed sampling, field and laboratory life tables, and only marginally modeling and theory. Often when applied ecologists develop models they do not examine the mathematical structure or properties.

Hence, it is not surprising that Levins and Wilson (1980) asserted that applied ecology lacked a theoretical basis, but this is not true. The methods summarized here are steeped in the theory of classical and theoretical population ecology. In the past, ecological theory was proposed but seldom tested except in ad hoc ways. This proposition led to an impasse between the extremes of population ecology. But this is not an unreasonable state of affairs, as such schisms have existed in all fields of science until a convincing theory develops to unify the discipline. This is beginning to happen in ecology as the extremes begin to recognize common ground (May 1973, 1981; Gilbert et al. 1976; Gutierrez and Wang 1976; Wang and Gutierrez 1980; Berryman 1992; Berryman et al. 1995; Gutierrez 1992; Gutierrez et al. 1994; and Arditi and Michalski 1995; Murdoch 1994).

May's books are important starting points for theoretical approaches, and the counterviews of Gilbert et al. (1976) are of historical interest (see Lawton 1977 for criticisms). Books by Varley et al. (1973), Hassell (1978), Pimm (1982), Price (1984), Walters (1986), Curry and Feldman (1987), Crawley (1992), Getz and Haight (1989), Royama (1992), DeAngelis et al. (1992), and hopefully this one make progress in closing the gap between theoretical and applied ecology.

SOME BASIC CONCEPTS

Realistic age-structured models of plants, herbivores and natural enemies, and of their tritrophic interactions are *time-varying life tables* (Gilbert et al. 1976), and their development is the focus of this book. As the name implies, time-varying life tables are not static. Rather, their parameters change over time and with factors such as age and the various biotic and abiotic factors we observe affecting species in nature. The use of time-varying life tables is increasingly common in biological control (the discipline) and its spin-off crop production and integrated pest management (CP/IPM). Both fields are aspects of applied population ecology, with the former having the special responsibility for the introduction and manipulation of natural enemies to control pests. CP/IPM integrates all factors that impinge on the agro-ecosystem with the objective of minimizing agro-technical inputs while maintaining high sustainable yields—it is an exercise in applied population ecology and ecosystem analysis and management with modeling serving as the unifying agent (Huffaker 1980). The basis of CP/IPM is natural and classical biological control.

The term biological control is often used to denote the regulation of a species at any trophic level by natural enemies (parasites, parasitoids, predators, pathogens, and antagonists) in a *self-sustaining manner*, though this definition has recently been eroded as more researchers jump on the bandwagon in vaguely related fields (see Garcia et al. 1988). Biological control, the phenomenon, has two components: natural biological control and classical biological control. Natural control is the biotic mortality that occurs in ecosystems without human intervention, and classical biological control is the purposeful introduction of natural enemies to control mostly introduced pests.

A CONCEPT OF PREDATION

Predation is the process of resource acquisition and allocation, and despite the simplicity of the concept, the terminology of prey–predator relationships is often cumbersome. The word **predator** is applied in its broadest sense to include plants that seek light, nutrients, and water, as well as herbivores (i.e., plant predators), true predators, parasitoids, parasites, and the human species, which is the consummate predator. The term **prey** will be applied to the victims of predation and may include light, nutrients, plant or plant parts, or animal victims. These notions are scattered in the literature, and will be adopted for convenience. Where necessary to clarify biological issues, precise terms shall be used.

THE CONCEPT OF DENSITY DEPENDENCE

Regulation of natural populations is thought to occur via the action of factors that operate in a density-dependent manner, and hence a brief exposé of this concept is justified. The concept of density dependence was first outlined by Smith (1935, 1939) though Howard and Fiske (1911) had identified salient components earlier. The types of possible interactions discussed here are depicted in Figure 1.1. Population interactions may be intraspecific or interspecific in nature. Interactions between members of the same species are intraspecific and normally take the form of competition for resources (food, mates, space, etc.). Interspecific interactions may be competition between members of different species for resources or predator–prey relationships. In general, if the effect of competition, predation, or other factors increases in severity as

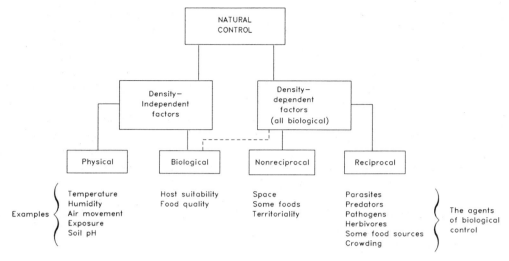

Figure 1.1. *Modes of action of biotic and abiotic factors in natural control.* [*cf., van den Bosch et al. 1982.*]

population density increases, the action is directly density dependent. For example, a predator acts in a density-dependent manner if the rate of kill of the prey increases or decreases as the density of the prey increases. If the effect decreases in severity with increasing density, the action is inversely density dependent. If the action is not correlated to density, such as the lethal effects of weather, the factor is termed density independent because it kills a constant proportion of the population regardless of density. However, the density dependent effect of a mortality factor operating in this generation or time step may not become evident until some future time, and such a factor is delayed density dependent if the effect can be shown to increase with density. Density dependent interactions tend to be reciprocal, but some are nonreciprocal. For example, individuals in a population may compete for a resource such as nesting sites, but the resource itself is not depleted—only its availability is affected and thus is nonreciprocal. Of course, abiotic factors may affect density dependent interactions, and in some cases may be the dominant factor determining the population dynamics of a species. These concepts have been the mainstay of biological control and population ecology for the last half century (see Huffaker et al. 1971) and are important arguments in Chapter 3.

The importance of density dependence in the regulation of populations has been challenged by Murdoch and colleagues (e.g., Murdoch 1991, Reeves and Murdoch 1985, 1986; Murdoch and Reeves 1987; Murdoch et al. 1985), who suggest a shift to the concept of stability in the context of the ebb and flow (*hide and seek*) between predator and prey in *metapopulations*. The concept of density dependence has also been cast in the jargon of systems theory—of positive and negative feedback (Levins 1975; Berryman 1981), of vague density dependence (Strong 1983), which I find lacks merit (see Berryman 1991), and more recently of ratio-dependent predation. Arditi and Ginsburg (1989) coined the term ratio-dependent predation, but the notion has been embedded in models for quite some time (see Thompson 1939; Gutierrez and Wang 1976; Getz 1984; Berryman et al. 1995). These ideas, however, remain within the ambit of the concept of density dependence. These different perspectives have much to do with scientific perspective, the units chosen to study the population interactions, failure to account for stochastic factors, spatial and temporal scales, and possibly the data and models used in the analyses (see Rosenheim et al. 1989; Wang and Gutierrez 1980). It is encouraging to see the incorporation of more biological realism in theoretical models and comparisons of model predictions to the real world (Southwood et al. 1989; Hassell et al. 1991; Murdoch 1994).

THE BASIS OF THE APPROACH

There are many modeling approaches to the study of multitrophic interactions. In general, most models used for this purpose have lacked age structure because to include it overly complicates the analysis, the motivation for predator search behavior has not been well specified, and the currency of flow between trophic levels has been numbers. Because of their simplified structure, such models often give unrealistic predictions, and hence have rarely been useful in resource management. This recognition provided ample reason for the development of an alternative paradigm to model plant–herbivore–natural enemy interactions for applied problems.

This alternative is one of *mass/energy flow dynamics* in an age-structured multitrophic population context. The plant level captures the light energy via photosynthesis that passes up the trophic chain, and this backdrop sets the stage for examining bottom-up as well as top-down effects on population regulation (see Hairston et al. 1969). The innovation is wedding physiological notions of per capita population growth (see Townsend and Calow 1981) with *classical population ecology*. The models describe the processes of energy acquisition and allocation across trophic levels in the face of observed weather and edaphic factors. The processes underlying these dynamics appear physiologically and behaviorally complicated and are known to be influenced by properties of individuals such as age, behavior, size, and other attributes. A recurring major theme of this book is that *common processes occur at all trophic levels*—a property that is exploited here. The identification of analogous processes for resource acquisition and allocation, and the similarity of the functions that describe them greatly simplify the problem making the development of realistic multitrophic models highly tractable.

Finally, the influence of economics is seen in our models in the use of the ratio of requisite *supply to demand* to regulate all birth, death, mass growth, and net emigration rates. This, as indicated by the title, is a core feature of this paradigm because this ratio captures the time varying behavior and dynamics of field populations. Factors either affect the demand or the supply side of the ratio. The physiology and behavior underlying the notions of supply–demand by-and-large enable our models to be independent of the population dynamics data they reproduce.

2

Sampling in Applied Population Ecology

If your experiment needs statistics, you ought to have done a better experiment . . .
—*Lord Rutherford*

Crucial elements of field ecology are experimental design and sampling. In this chapter, aspects of sampling that have made major contributions in applied ecology and that have relevance to our modeling paradigm are examined.

A POINT OF VIEW

Scientists are trained to test hypotheses, to look for differences, and yet when we are faced with the complexity of *community and field population ecology*, formulating comprehensive hypotheses proves difficult except in abstract ways (e.g., $r - K$ selection and optimal foraging, simple models of population dynamics). Such exercises may be academically interesting, but they may be difficult to relate to systems in nature. This situation is especially true where there is species richness, the species reproduce all year round, the generations overlap greatly, and the details of the biology are complicated. Ecologists may be so overwhelmed by the perceived complexity of an ecosystem that many satisfy themselves examining isolated bits of the puzzle. But to do small set piece experiments in complicated ecosystems because they can be handled with the hypothesis testing paradigm is, in my opinion, not very satisfying or useful if the task at hand is the management of an ecosystem, a population, or a renewable resource.

Frequently, the only step taken in the analysis of the field data is to plot them against time. This approach has a long tradition in the literature, but often contributes little more than summary raw data to our understanding of events in the field. The use of ANOVA, regression analyses, Markov chain models, or other statistical models to predict population size or the factors affecting it may prove an unsatisfactory exercise (see Morris 1963; Varley et al. 1973), often, because we may not know the appropriate relationships, they are likely to be nonlinear in any case, and may involve time lags and largely unknown stochastic factors. The analysis may be useful for estimating net effects over a season, but not for predicting within season dynamics.

Such analyses may be extremely useful for examining factor interactions on the yield in field survey data. For example, Neuenschwander et al. (1989) successfully used econometric applications of multiple regression analysis on extensive survey data to estimate the effects of several biotic and abiotic factors on cassava yields. They were able to document the reduction of cassava mealybug populations (*Phenacoccus manihoti* Matile Ferrero) by the introduced hymenopterous parasitoid *Epidinocarsis lopezi* DeSantis, and the resultant increase of 2 metric tons of cassava per hectare across the humid subtropical African Savannah zone. Their methods could not, however, explain the underlying dynamics of the interacting species and weather—this required the development of a realistic tritrophic population dynamics model that reproduced and explained the field data (Gutierrez et al. 1993). To develop this model required that the dynamics data for all species be accurate and complete. Figure 2.1 summarized the types of models used for different research emphases and applications (cf. Baumgärtner et al. 1987).

In ecology, we must be prepared to paint with a broad brush because little may be known about the interaction of the species in the system—to fill in the details later. A common presumption is that the complexity of an ecosystem precludes all but enlightened guesses as to the factors governing its dynamics. This is due to the fact that the theory underlying field ecology, as opposed to the physical sciences, is poorly defined. We are often forced to rely on our intuition about the system in formulating research questions and models about the dynamics of species.

In any ecological study, it is important to first identify the goals of the study, the relevant species (*taxonomy*) to be studied, and, if modeling is a component, the form of the model. We need to determine which aspects of the biologies are known and which have large gaps. As obvious as these steps might seem, they are often neglected—all too often the researcher simply begins collecting data. When beginning an ecosystem study (as opposed to hypothesis testing), it is well to first think of all of the reasons why the study is not worth pursuing as it may prove difficult, time consuming, and costly, and there are many blind ecological alleys that the incautious might fruitlessly pursue. If the system is complicated, we might be unable to sketch the biological relationships, and the first field studies might be designed to simply enable us to do this. The decisions as to how much of the system may be safely ignored must be made. This decision requires considerable judgment—there is no substitute for field experience. Finally, experience on a wide range of ecosystem studies has shown

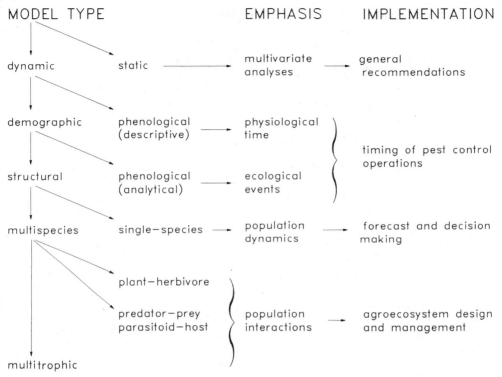

Figure 2.1. *A diagram of model types, research emphases, and applications. [Redrawn from Baumgärtner et al. 1987.]*

that the data required to develop a model are considerably less than those required to test it.

SAMPLING

Data in the literature are often reported in reduced form. In the worse case, they may be sparse, poorly gathered or summarized, or worse yet, the experiments to gather them may have been ill conceived and poorly executed. Incomplete data may give us false insights. Laboratory data and field observations of highly competent ecologists, however, remain as valid today as when they were collected [e.g., Huffaker's (1958) orange mite data, Nicholson's (1954a,b; 1957) blowfly data, and Isley's (1932) boll weevil data], but such data are rare and even they are not entirely above suspicion.

Once the decision has been made to pursue a research problem, the design of the approach and the sampling scheme itself becomes paramount. Should one first study the pieces separately or should a broader approach be taken? In general, I have found that it is best to study simultaneously as many of the relationships in the same field as possible—it makes cross checking of our developing understandings easier and enhances the possibility of noticing unforeseen interactions. Of course, the process of collecting the data may simplify the problem, or may show that an increased effort is required in certain areas because components may not have been adequately considered when the study was designed. One can study the same system simultaneously under different climatic regimes by sampling other localities—this is akin to studying the system at the same locations over several years. The notion that we must study the system several years to be able to model it is part of the common wisdom, and may become a self-fulfilling prophesy.

It is important to sample as extensively and as often as humanly possible, as this avoids placing too much emphasis on point estimates that may prove to differ from the apparent trend due to sampling error alone. This is called the *comprehensive approach*. An alternative is to study components of the system separately in different experiments over a long period of time, and then hope the pieces of the ecological puzzle come together—this is called the *linear approach*. The latter approach often proves unsatisfactory as different field experiments have different uncontrolled variables that makes melding of the data sets nearly impossible, except by grand extrapolation. In physics and chemistry, the repeatability of an experiment is more of a foregone conclusion because the conditions are often highly controlled and the laws underpinning the science are based on deep theory. This is not the case in field ecology where biotic and abiotic factors introduce considerable variability, which may propagate over the course of the experiment, and the theory often does not survive a scratch to its surface.

There are three common considerations in sampling: (1) what data to collect, (2) how to collect the data, and (3) at what level of precision. As obvious as this might seem, all too often the collection of field data is a hit or miss proposition. Here we emphasize field sampling that will be useful for testing the predictions of the models and paradigm outlined in this book. For this purpose, laboratory and field work must go hand in hand. The aim here is not to develop an exhaustive protocol, but rather to point the direction to gathering the minimum but sufficient data set.

WHAT DATA TO COLLECT AND HOW TO COLLECT THEM

Field data should, among other things, estimate total density defined as the numbers in all stages per biologically meaningful unit. The vagueness of the term unit is needed because the biology, and at times the physical constraints, of sampling may determine sampling unit size. Different species operate on different spatial and temporal scales. The sampling unit appropriate for one may be inappropriate for another. Judgment needs to be exercised in the selection of the sampling unit, but we must recognize that compromises often have to be made. Age, mass, morph, and sex structure of populations should also be assessed because much of the dynamics we will wish to model and analyze concern such issues. The trade-off between sampling accuracy and the effort required to achieve it is well known, hence it is best to recognize this compromise beforehand rather than later when one is trying to reconcile discrepancies in field data—especially when one is trying to explain patterns in the data.

So-called *absolute sampling methods* may provide accurate estimates for most species in the sampling unit, but they are often very time consuming to process. An example of an absolute method is the whole plant bag sampling method (WPBS) used by Byerly et al. (1978) and others to sample arthropods in cotton (*Gossypium hirsutum* L.). The method consists of placing organdy sleeves over each of several plants (it could also be a branch), securing them at the bottom and collapsing each at the base of the plant. Several days later, after the populations of mobile organisms have reestablished, the sleeves are quickly pulled over the plant and sealed at the top, the plants are cut at the base and taken to the laboratory where all organisms trapped in the sleeve are counted. The plants themselves can be dissected and their subunits counted, dried and weighed, and the data used to develop models for plant dynamics. Although absolute methods may be accurate, they may not prove feasible for very large plants, when sampling overly depletes the population or when time is limiting.

Relative methods include samples obtained using light traps, pheromone traps, sweep nets, suction devices, abundance indices, and other such methods. In general, these estimates are not very satisfactory except for phenological studies. If the sampling efficiency of a relative method is unknown, rather severe problems may arise when the estimate is used in decision making. For example, during the period 1960–1975, *Lygus* bug (*Lygus hesperus* Knight) was considered the key pest in the 1 million acres of cotton in the San Joaquin Valley of California (Falcon et al. 1971), and an economic threshold (i.e., *the point at which the value of the yield loss is equal the cost of the control*) of 10 bugs per 50 sweeps was used to determine when a field should be treated. Approximately \$20–40 million was spent annually on insecticides to control *Lygus* bug and its resurgence and secondary pests induced when pesticides killed natural enemies (van den Bosch 1978). Byerly et al. (1978), using the WPBS method, showed that the sweepnet method captured less than 1 in 12 plant bug nymphs and only 1 in 3 adults. The seriousness of the sampling error is obvious, but the folly of an ill-founded economic threshold based on an inaccurate sampling method was compounded by an inadequate understanding of the plant—herbivore dynamics. Despite a presumed need for control, yields were found to be consistently higher in untreated than in treated cotton (Falcon et al. 1971; Gutierrez et al. 1975). *Farmers had been spending money to lose money, clearly demonstrating market failure in cotton pest control in this area.*

Ecological studies, accurate sampling methods, and the development of models of the kind outlined in this text explained this apparent paradox (Gutierrez et al. 1977b)—the per capita damage rate due to *Lygus* bug was smaller than presumed by entomologists and farmers. *Lygus* did not prefer cotton (Cave and Gutierrez 1983) and the plant could compensate for the damage caused by the small populations of this pest, which are common in cotton (Gutierrez et al. 1977b). This example shows not only the sampling errors that may be involved, but also the need to understand the nature of the underlying dynamics.

It is important to determine the within-plant (within field) distribution of each species, as this helps determine the appropriate sampling unit and may possibly lead to the discovery of relevant relationships. For example, a predacious species found mostly on the top of the plant may not interact strongly with prey species at the bottom, unless there is movement not accounted for by our sampling (e.g., Pizzamiglio et al. 1989). This cautions against an uncritical use of correlations that may show that the two species are temporally synchronized, where they are in fact not responding to each other, but rather to other factors.

Population densities of each species should be estimated as frequently as possible as this may enable us to distinguish sampling error from the effects of fast-acting mortality factors. A sharp decline in a population may occur due to, say, heavy rainfall, and the effect might be missed if one samples infrequently. As logical as this admonition might seem, it is a recurring problem in data sets.

Estimates of average per capita dry biomass of the different stages of each species should be made frequently during the season. Specimens may be put in alcohol or frozen for later weighing if time does not permit immediate processing of the samples. As we shall see, size data compared to the maximum provide estimates of the favorability of environmental conditions for the population over time. Organisms that have been stressed due to poor nutrition or adverse weather are smaller and their survivorship and fecundity are reduced. Data on size may enable us to separate these effects on population growth rates from, say, those due to natural enemies. We shall later see how this is done.

Dissection of fresh samples to determine the condition of the host (number of ova, diapause status, etc.), parasitism rates, super- and multiple parasitism, disease incidence, host preference, and other factors is preferable to all alternatives. Dissection of specimens preserved in alcohol or other fixatives is often vexing and the results generally less satisfactory. Of course, many of these factors can be estimated by rearing field samples, but much information may be lost.

Sample sizes should be as large as resources allow, unless we have the luxury of developing a sampling plan first and then doing the ecological research later. Often we save time by initially doing intensive sampling, and developing sampling rules from the data. The analysis can tell us whether the sample size was adequate or whether we need to increase it or can safely decrease it. This problem is explored below.

LEVEL OF PRECISION OF FIELD DATA

There is an extensive literature on this subject, and only some of the relevant aspects are reviewed here. Karandinos (1976) stimulated considerable interest in developing sampling decision rules to estimate the optimal sample size, and is the basis for this section. Nachman (1982) made some important contributions to the development of binomial sampling decision rules that are reviewed by example. Ruesink and Kogan (1982) and Wilson and Room (1982) extended Karandinos' work to include the so-called Taylor's power law (Taylor 1961), describing the relationship between the variance and mean. Both groups distinguish sampling to estimate population mean density from sampling for Integrated Pest Management (IPM) decision making. In the latter case, one wishes to know with a predetermined level of precision whether the population mean is above the economic threshold. These two problems are illustrated in Figure 2.2 (cf. Wilson et al. 1989), and are discussed below with an eye cast to real

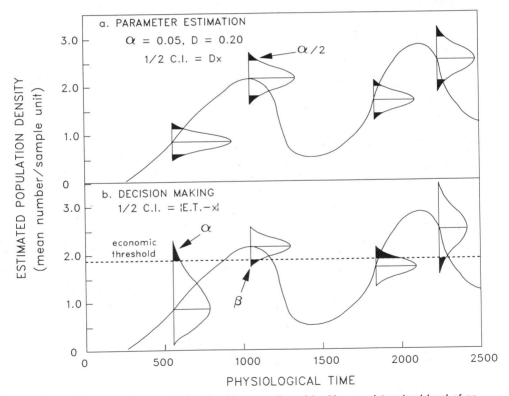

Figure 2.2. *Estimating population densities over time: (a) with a predetermined level of accuracy, and (b) for decision making using an economic threshold (ET). [Redrawn with permission from Wilson et al. 1989.]*

world problems. It is commonly accepted that the economic threshold is not static, it is a complex function that may vary over the season.

PRECISION OF THE ESTIMATE

A major concern in population dynamics studies is the accuracy of the sample mean ($\overline{X}$) for predicting the true population mean (μ) with variance σ^2, where $x_1, \ldots, x_n$ are observations in a random sample.

$$\overline{X} = \sum_{i=1}^{n} x_i \Big/ n \qquad [1]$$

In practice, the sample variance s^2 is substituted for σ^2 and $\overline{X}$ for μ in these analyses. The coefficient of variation (c.v. $= s/\overline{X}$) is used to estimate the level of precision [D] of the random variable x

$$D = \frac{s}{\sqrt{n}} \Big/ \overline{X} \qquad [2]$$

This implies that as sample size n increases, the c.v. decreases. Say that we wish a level of precision (i.e., $D = 0.1$), then we can solve [2] for n as follows:

$$n = (s/D\overline{X})^2 \qquad [3]$$

Karandinos points out that . . . *to design a good sampling strategy and specifically determine the optimal sample size, we must know the distribution of the random variable x* . . . . For example, if the parent distribution is a negative binomial, then its variance may be described by

$$s^2 = \overline{X} + \frac{\overline{X}^2}{k} \qquad [4]$$

where k is the parameter of the negative binomial distribution and the variance is larger than the mean. Substituting [4] in [3] yields the number of samples n required to meet the level of precision D:

$$n = \frac{(1/\overline{X}) + (1/k)}{D^2} \qquad [5]$$

Karandinos cautions that the reliability of the estimate using [5] may not be high.

Similarly, because in the Poisson distribution $\mu = \sigma^2$, an appropriate substitution may be made in [3] yielding

$$n = \frac{1}{D^2 \overline{X}} \qquad [6]$$

For binomial distributions (presence–absence phenomena), we are interested in the probability (p) of something being present in one trial. This index is estimated from x of n trials or observations as

$$p = \frac{x}{n} \qquad [7]$$

with variance of p defined by Freund (1962) and Freund and Walpole (1987) as

$$s_p^2 = \frac{pq}{n} \qquad [8]$$

Note that the probability of it being absent in n trials is $q = 1 - p$. Therefore, if

$$\text{c.v.}(p) = \frac{s_p}{p} = D \qquad \text{and} \qquad s_p = Dp \qquad [9]$$

then substituting s_p from [9] in [8] and solving for n yields

$$s_p^2 = (Dp)^2 = \frac{pq}{n} \qquad \text{and} \qquad n = \frac{q}{pD^2} \qquad [10]$$

For sufficiently large samples, the *central limit theorem* gives the probability statement

$$\text{prob}\left\{ \overline{X} - z_{\alpha/2}\left(\frac{s}{\sqrt{n}}\right) < \mu < \overline{X} + z_{\alpha/2}\left(\frac{s}{\sqrt{n}}\right) \right\} \approx (1 - \alpha) \qquad [11]$$

where $(1 - \alpha) = 0.95$ is the confidence coefficient, and $z_{\alpha/2}$ is the upper $\alpha/2$ of the standard normal distribution. The confidence interval (C.I., [11]) can be expressed as a proportion (D) of the mean

$$D\overline{X} = \frac{s}{\sqrt{n}} z_{\alpha/2} \qquad [12]$$

Solving [12] for n gives the general formula for predicting the number of samples required for a predetermined level of precision D (see Southwood

1975) for the sample values

$$n = D^{-2} z_{\alpha/2}^2 \ (s^2 / \overline{X}^2)$$ [13]

TAYLOR'S POWER LAW

Equation [13] has been used by Ruesink and Kogan (1982) and Wilson and Room (1982) as the starting point for developing sampling decision rules that utilize Taylor's (1961) power relationship [14] relating the mean and the variance across different statistical distributions.

$$s^2 = a\overline{X}^b$$ [14]

This model describes a wide range of distributions: $a = b = 1$ implies a Poisson distribution, $a < 1$ and $b > 1$ imply a clumped distribution, and $a > 1$ and $b < 1$ imply an under dispersed or uniform distribution. More precisely, a clumped distribution is characterized by $s^2 > \overline{X}$ and is met in the Taylor power law when $a\overline{X}^b > \overline{X}$ or $a\overline{X}^{b-1} > 1$. Hence, when $\overline{X} > (1/a)^{1/b-1}$ the population is clumped and when $\overline{X} < (1/a)^{1/b-1}$ the population is underdispersed. The conditions for clumpedness are unlikely to be met for $a > 1$ and $b > 1$.

Others, Fracker and Brischle (1944) and Hayman and Lowe (1961) found this power relationship before Taylor (see Wilson et al. 1989), but Taylor popularized it and in the literature it has assumed his name. The coefficients a and b are easily estimated by regressing log s^2 on log $\overline{X}$.

The numbers of samples generalized across distributions for different levels of accuracy D defined as the proportion of the mean, can be obtained from Karandino's formula by substituting [14] in [13]

$$n = Z_{\alpha/2}^2 D^{-2} a\overline{X}^{b-2}$$ [15]

Although not based on the Taylor power law, the equivalent substitution for binomial or presence–absence data is

$$n = Z_{\alpha/2}^2 D_p^{-2} p^{-1} q$$ [16]

where p is the proportion of the sampling units infested, q is the proportion not infested, $D_p = CI/2p$ is the level of precision, and CI is the confidence interval (see [11]).

APPLICATIONS

Numerous studies utilizing Taylor's method and adaptations of the method have appeared in the literature, and my selection of two Brazilian data sets is made

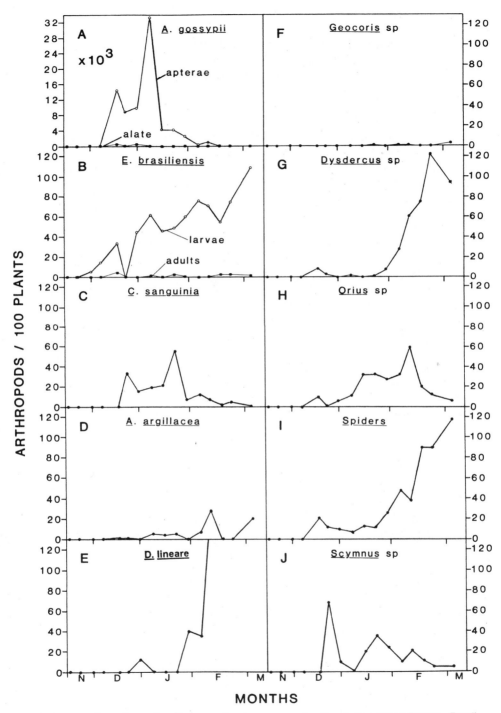

Figure 2.3. *Phenology of various species in unsprayed cotton in Londrina, Parana, Brazil based on 100 whole plant bag samples. [Pizzamiglio et al. 1989.]*

because they are very extensive, convenient, and illustrate two different sampling objectives, namely, estimating the mean and developing easy to use presence–absence decision rules for pest control.

Estimating the Mean

In the first case, Pizzamiglio et al. (1989) wished to establish the expected phenologies of common economically important insect species found in unsprayed cotton prior to the arrival of the bollweevil (*Anthonomus grandis* Boh.) in Southern Brazil. The data serve as a baseline to determine the degree of ecological disruptions caused by increased pesticide use for control of this pest. A sampling method was needed that gave roughly the same level of accuracy for all species, hence 100 whole plant bag samples (WPBS) were laboriously processed per sampling date (Fig. 2.3). To develop the sampling decision rule for each species, mean density and variance for each species on all sampling dates were computed, and a simple log linear regression of s^2 on $\overline{X}$ was used to estimate the species specific intercept a and slope b of the Taylor power law (Table 1). The number of WPBS units required to sample each species at different densities for the three levels of precision were computed using decision rule [15] (Fig. 2.4).

Based on the interpretation of a and b, *Geocoris* sp. was greatly underdispersed at the whole plant level, the cotton stem borer [*Eutinobothrus brasiliensis* (Hambelton)] was randomly distributed, and many of the other species exhibited varying degrees of clumping with the cotton aphid (*Aphis gossypii* Glover) exhibiting the highest degree. As expected, the results using [15] show

TABLE 1. Fitted Parameters for Taylor's Variance—Mean Power Relationship (Natural Logarithms): Whole Bag Plant Sampling Data from Cotton at Londrina, PR, Brazil During the 1982–1983 Season[a]

Species	N	a	b	r^2
E. brasliensis adults	9	1.116	1.041	0.99
Larvae	13	1.029	0.903	0.94
C. sanguinea	12	2.461	1.253	0.98
A. argilacea	8	2.375	1.247	0.94
Dysdercus sp.	10	0.883	1.912	0.75
D. lineare	8	2.655	1.139	0.96
E. conexa	8	1.538	1.093	0.97
Geocoris sp.	5	0.046	0.132	0.66
Orius sp.	12	1.356	1.048	0.97
Scymnus sp.	12	2.145	1.230	0.98
spiders	12	1.703	1.052	0.86
A. gossypii	13	1.855	1.741	0.98

[a]Pizzamiglio et al. (1989).
N is number of sampling periods.
a and b are coefficients of Taylor's power law.
r^2 is the coefficient of determination.

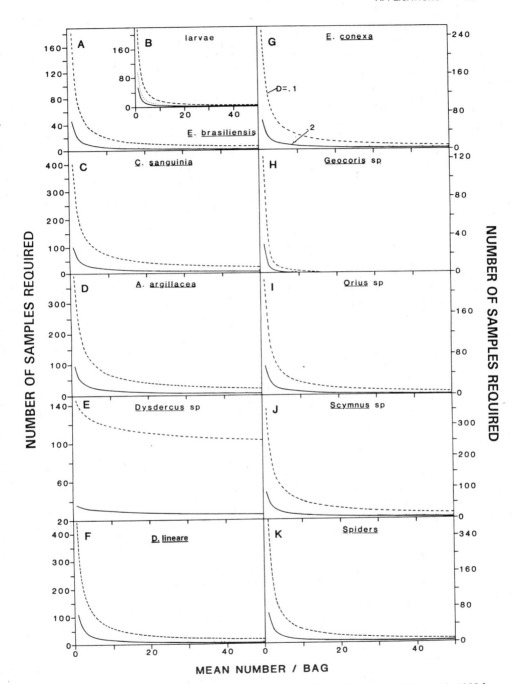

Figure 2.4. *Sampling decision rules for the species in Figure 2.3. [Pizzamiglio et al. 1989.] The definitions for the different levels of precision are given in the text.*

that fewer samples would be required for level of accuracy $D_p = 0.2$ than for $D_p = 0.1$ (Fig. 2.4). We should note that the Taylor parameters are highly dependent on the sampling unit and the method.

PRESENCE–ABSENCE METHODS FOR DECISION MAKING

In the second study, Villacorta and Gutierrez (1989) sought to establish sampling rules for the coffee leaf miner (*Leucoptera coffeella* G-M) to determine when insecticides should be applied for control when populations rose above a *static economic damage threshold* of one leaf miner lesion per leaf. In fact, such thresholds are known to vary widely with previous damage, age and size of the plant, nutrient levels, and so on. A sample of 100 leaves were taken throughout the season in treated and control blocks, and the data were summarized in Figures 2.5a. The effort required to gather these data was relatively small compared to the WPBS method used for sampling cotton insects. The sampling decision rule for the numerical data (lesions per leaf) is shown in Figure 2.5b for $D = 0.1$ and 0.2. The results suggest that a 250 leaf sample needs to be taken when $D = 0.1$ to accurately estimate lesion density at the economic threshold (solid line), but less than one-half that number are required for a level of precision $D = 0.2$ (dotted line).

Presence–Absence Sampling Method

In IPM, however, we often want a very quick but accurate method for estimating the population mean to decide if the population is above or below the economic threshold. The easiest method to use is a *presence–absence* or *binomial sampling* method. From this point of view, the decision rules for the number of coffee leaves needed to estimate lesion densities at different levels of precision may be computed using model [16]. The data points superimposed on the line (see Fig. 2.5c) were estimated by substituting specific observed pairs of p and $\overline{X}$ in [16].

We may also predict lesion density from the proportion of leaves infested (p), say at $D = 0.1$. To do this we would first plot the proportion of leaves infested on observed lesion density (Fig. 2.5d). We should recognize that the level of reliability of the decision rule is not equivalent for both p and $\overline{X}$ (i.e., $D_p < D_x$), since the number of samples required to assure the same level of accuracy increases as p asymptotically approaches unity (Wilson et al. 1989). In general, more samples need to be taken to correct for this asymmetry, and this is indicated by the dashed line in Figure 2.5c. The simplest model that might fit the data on the proportion of leaves infested on mean density per leaf (Fig. 2.5d) is the Poisson model (i.e., the dashed line)

$$p = 1 - e^{-\overline{X}} \qquad [17]$$

where $e^{-\overline{X}}$ is the zero term of the Poisson distribution and $\overline{X}$ is the mean lesion

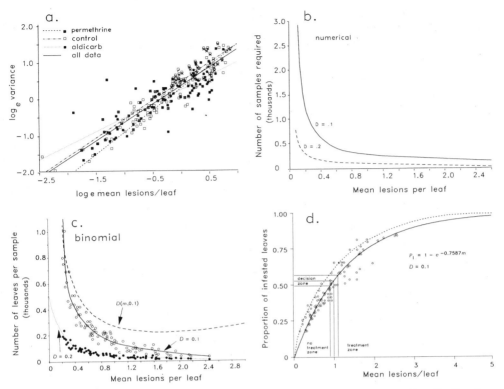

Figure 2.5. *Coffee leaf miner (*Leucoptera coffeella *G-M) lesion data from Londrina, Pr., Brazil (cf. Villacorta and Gutierrez 1989): (a)* $\log_e$ *variance on* $\log_e$ *lesions per leaf in three chemical treatments and a check, (b) numerical sampling plan for levels of accuracy* $D = 0.1$ *and 0.2, (c) a binomial sampling plan with constant levels of precision* $D = 0.1$ *and 0.2 as well as a sampling plan where the level of precision varies with mean lesions per leaf (see Wilson and Room 1982), and (d) a sampling decision rule based on the proportion of infested leaves (the dashed line is the Poisson model and the solid line is a modified Poisson model—see text).*

per leaf. This model, however, is unsatisfactory, and a better fit is obtained using a modified Poisson model (the solid line):

$$p = 1 - e^{-0.76\overline{X}} \qquad [18]$$

Unfortunately, [18] lacks a sound theoretical basis and the result is at best an exercise in curve fitting.

Wilson and Room (1982) propose a "general model" for predicting p that includes the Taylor power relationship [19].

$$p = 1 - \exp\left[-\overline{X} \log(a\overline{X}^{b-1})/(a\overline{X}^{b-1} - 1)\right] \qquad [19]$$

However, because of the complexity of this model, readers are advised to consult Wilson et al. (1989) for details of its derivation. A simpler model

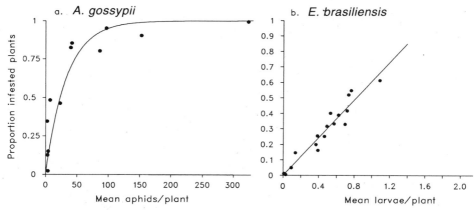

Figure 2.6. *Plots of the proportion of cotton plants infested on mean infestation level per plant: (a)* A. gossypii *and (b)* E. brasiliensis. *[Pizzamiglio et al. 1989.]*

proposed by Nachman (1984) has great utility for presence–absence data

$$p = 1 - e^{-a'\overline{X}^{b'}}$$ [20]

where a' and b' are constants.

We could fit [19] or [20] to presence–absence data and superimpose an economic threshold level to decide whether to spray or not. For example, using the threshold of one lesion per leaf, Villacorta and Gutierrez (1989) developed a *treat and no treat decision rule* for coffee (Fig. 2.5d). When the predicted $\overline{X}$ falls in the decision zone, the sample is insufficient to determine whether the infestation is high enough to treat. In practice, most farmers would probably treat in this case, and despite this ambiguity many treatments would be avoided simply because an objective method was used to make the decision.

Similar binomial or presence–absence sampling decision rules were developed for other highly visible species such as the cotton aphid and the cotton stem borer studied by Pizzamiglio et al. (1989) (Fig. 2.6). For cotton aphid, the results may have little practical value as the proportion of infested plants saturates to unity at densities far below the economic threshold and the variance may be very large. In contrast, a linear relationship was found for the highly destructive cotton stem borer Fig. 2.6b. The economic threshold for this pest is approximately 0.2 larvae per plant, and this is well within the valid linear range of the model (Dos Santos et al. 1989).

SEQUENTIAL SAMPLING USING BINOMIAL DATA

Sequential sampling decision rules developed from presence–absence data may be used to reduce the sampling effort further. A presence–absence sampling plan proposed by Wilson et al. (1983) was used by Bianchi et al. (1989) to develop a sequential sampling plan for egg masses of the lepidopterous African white stem borer (*Maliarpha separatella* Rag.) on rice. The Wilson et al. (1983) model is

$$n = (z_{\alpha/2})^2 (p - T_a)^{-2} p(1 - p) \qquad [21]$$

where p is the proportion of infested units in n samples and T_a is the action threshold equal to the confidence interval defined as a statistical statement with $z_{\alpha/2}$ being the upper part ($\alpha/2$) of the standard normal distribution with a confidence coefficient $(1 - \alpha) = 0.95$ (see Karandinos 1976). As Bianchi et al. (1989) point out, the confidence coefficient is often selected for pragmatic reasons.

Bianchi et al. (1989) used Nachman's (1984) model [20] to predict the injury level from the proportion infested. The ratio of the number of infested sampling units x among n samples (x/n) is substituted for p in the Wilson et al. (1983) model [21]. By noting that $z_{\alpha/2}$ and T_a are constants and rearranging [21], a quadratic equation for x is obtained with solutions x_1 and x_2 and used to determine the upper and lower decision lines as a function of sample size n_s (Fig. 2.7).

To use this decision rule in practice, one starts with an initial small sample size n_s larger than some minimum size n^*. The number infested n_{inf} is determined, and if the point (n_{inf}, n_s) falls within the shaded area, additional sample units are taken up to a maximum of n for a given level of precision D determined using [15] for a specified $z_{\alpha/2}$ (Bianchi et al. 1989). In practice, one would continue taking samples until the number of infested units in samples falls

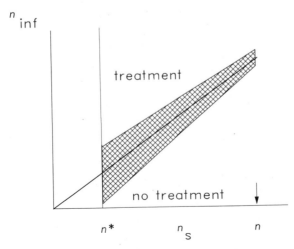

Figure 2.7. Hypothetical sequential sampling decision rule for presence–absence data. The shades area is the no decision zone, n_s is the number of samples taken, n_{inf} is the number infested, n^* is the minimum number of samples taken, and n is the maximum for a predetermined level of precision (see text). [Redrawn from Bianchi et al. 1989.]

outside of the shaded area or n_s equals the number of samples n for the specified level of precision [21].

EPILOGUE

One would suppose that collecting the appropriate field and laboratory data required to develop and test a multitrophic population model would take several years and a very large field staff, but this need not be the case. During the summer of 1971, N. E. Gilbert and I studied the difficult thimbleberry–aphid–natural enemy system on the campus of the University of British Columbia for the first time. The complexity of the biology of the holocyclic aphid [*Masonaphis maxima* (Mason)] and its parasitoids and predators at first seemed daunting, but it was quickly unraveled and a model was formulated. What we could not develop posthaste was confidence in the model—that required additional seasons of work. More data are required to test the model than to formulate it.

But how much data do we need to collect to model a system and to make sure that we understand it? Our work in cotton provides some insights into this question. In 1972, very extensive field data sets on cotton growth and development and on arthropod population dynamics were collected by Paul Leipzig and me (both then young and novices in cotton research). We gathered data on the effects of plant density on growth rates, we mapped the branching patterns of the plant, recorded the age and position of all leaves, the proportion of each leaf eaten, the presence of all organisms by species and age on each leaf, and finally, we dissected the plant by branch and dried and weighed the component parts. Other data were collected and the numerous cotton references cited in the bibliography should be consulted for details. The important point is that aside from the drying process, the whole exercise required less than 2 minutes per plant, some with 30+ branches, because we had developed a simple code beforehand, which allowed one person to read the plant while the other recorded the observations. These data remain the cornerstone of our cotton modeling effort today; this, despite the fact that more that 20 other data sets of equal quality from various regions of the world have been gathered to improve and test the model. Each data set gave different dynamical patterns, but the underlying mechanism explaining them proved deceptively simple and the same—a paradigm of how organisms acquire and allocate resources. This paradigm has served us well in modeling such diverse systems as Australian pastures (Gutierrez et al. 1971, 1974a,b), alfalfa (Gutierrez et al. 1976; Gutierrez and Baumgärtner 1984a,b; Gutierrez et al. 1984), cotton (Gutierrez and Curry 1989; Gutierrez et al. 1975, 1977a,b, 1984), grape (Gutierrez et al. 1985), cassava (Gutierrez et al. 1988a–c) and common beans (Gutierrez et al. 1993). The approach has also been used by colleagues and students to model still other systems (apple, Baumgärtner et al. 1986a; grape, Wermelinger et al. 1991a,b; rice, Graf et al. 1990a,b; Bianchi et al. 1990; cowpea, Támo and Baumgärtner 1993). It is the major theme of this book.

3

The Role of
Abiotic Factors

*The fundamental niche of a species is defined in terms of the abiotic environment—
the interactions with biotic agents defines the realized niche.*
—cf., G. E. Hutchinson

The importance of weather on the dynamics of animal (and plant) populations is aptly illustrated by reference to a mostly Australian controversy in population ecology. The controversy was whether weather (and climate in the long run) (Andrewartha and Birch 1954) or density dependent factors (Nicholson 1933) regulated population numbers. In hindsight, it is easy to see why each held strong views on the matter. Andrewartha and Birch studied field populations under highly variable Australian conditions, and their evidence suggested that numbers were controlled by extremes of weather. In contrast, Nicholson studied sheep blowfly populations under laboratory conditions and was impressed with the regular cyclicity of their numbers. The similarity to the observed cyclicity of other species in nature (red fox, lynx, and hares, Elton 1939, 1942) served to strengthen the view of many early ecologists that biotic factors operating through density dependent processes were responsible for these patterns.

We now know that both camps had important parts of the ecological puzzle— that biotic and abiotic factors interact to determine the dynamics of species growth and development—density dependent factors operate within the constraints set by weather and other physical factors. **Weather** is defined as the current set of conditions experienced by a population at a specific time and place, and among other factors helps set the limits for the growth of a population. In contrast, **climate** is the long-term historical pattern of weather that with other abiotic and biotic factors ultimately sets the limits for the distribution of the species, and is the background upon which natural selection operates.

In the absence of natural enemies and competition, plant (or animal) growth may be limited by a single nutrient in short supply (Liebig's law of the minimum, 1840), but realistically it is reduced or possibly limited by the compounded partial shortfalls of all resources (see Gutierrez et al. 1988b). The limits of favorability with respect to abiotic factors is Hutchinson's (1959) concept of the fundamental niche (the maximum limit of abiotic conditions favorable to population growth, see Fig. 3.1). The actual niche may be smaller because biotic factors such as competition together with abiotic factors determine the geographic distribution of a species in nature. Evolution plays a role in determining the precise parameters of species (White et al. 1970, Taylor 1981). Some factors such as temperature, humidity, rainfall, wind, and light intensity may fluctuate widely over relatively short periods of time and affect organisms directly, while factors such as photoperiod and soil pH, nitrogen concentration, and water availability may change relatively slowly. These factors (e.g., soil nitrogen) may affect plants directly, and indirectly the herbivores feed on them and higher trophic levels as well.

The rates of development of all plants and most invertebrate animals vary with the temperature of their immediate environment (i.e., **ectothermic or poikilotherms**). Most vertebrate animals control their body temperature using metabolic energy (i.e., **homeotherms**). The rate of development of all organisms may vary with the availability of essential resources (e.g., nutrients and water for plants and food quantity and quality in animals).

This chapter emphasizes the role of temperature and nutrients on the rate of poikilotherm development. A wide variety of behaviors may also play important roles in temperature regulation in invertebrates such as insects. For example,

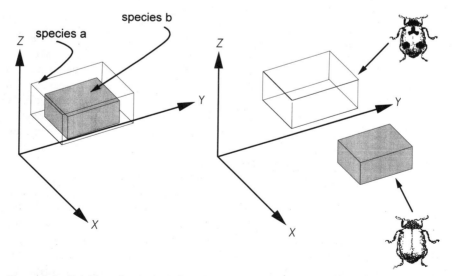

Figure 3.1. *Hutchinson's concept of the fundamental niche: The left figure depicts niche overlap and niche separation on factors X, Y, Z.*

some butterflies position their wings to absorb or avoid light, and dark insects, such as the palearctic cereal leaf beetle (*Oulema melanopus* L.), crawl to the top of plants on cool days and orient as black bodies to capture radiant energy to speed up their development. Behaviors that increase body heat may be beneficial in north temperate latitudes, while behaviors that dissipate heat may evolve in warmer climes. The physiology of dormancy (i.e., periods of physiological quiescence) evolved to enable some organisms to survive adverse periods, and hence plays an important role in life cycles of many plants and animals.

The microenvironment of small organisms (plants and animals) may differ from the ambient conditions, and obvious examples are aquatic species, and those that live in the soil or that mine through their host. Models of the effects of environmental factors may be empirical or biophysical. An example of the latter is the elegant biophysical microclimate model developed by Baumgärtner and Severini (1987) to estimate the temperature experienced by larvae of the leaf miner *Phyllonorycter blancardella* F as they burrow in apple leaves.

These and other mechanisms and factors used to regulate metabolic and developmental rates are important components that affect the population dynamics of species and their interaction. They may need to be included in models if the full dynamics are to be accurately characterized.

EFFECTS OF TEMPERATURE ON POIKILOTHERM DEVELOPMENTAL RATES

There is a large literature on attempts to transfer the notions of the rate of chemical reactions to model the developmental rates of poikilotherms. This transition is best made by noting that in general, the developmental time (T) in days of a poikilothermic organism is longer at low temperatures than at high temperatures (Fig. 3.2). If we were to run an experiment where cohorts of poikilotherms were placed at different temperatures with all other factors held constant, a plot of the developmental times on temperature (τ) would be monotonically decreasing. A plot of the reciprocal of developmental time [i.e., the rate of development $T(\tau)^{-1} = R(\tau)$] on τ yields a sigmoidal function (Fig. 3.2a). The rate of development on τ is roughly linear in the middle of the range and increases at a decreasing rate to a maximum.

Both linear and nonlinear models have been used to characterize this relationship (Fig. 3.2b). Huffaker (1944) and Gordon (1984) summarized the early work on the effects of temperature on poikilotherm rates of development and Curry and Feldman (1987) review the recent literature.

Linear Models

Often, data on development across the complete range of temperatures (τ) are not available, making the linear model a convenient starting point.

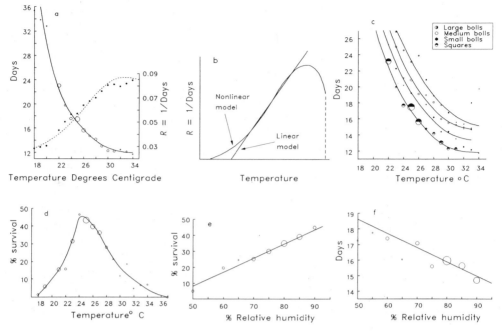

Figure 3.2. *The effects of food, temperature, and humidity on cotton boll weevil: (a) the effects of temperature on developmental time and rate, (b) a linear and nonlinear model of the rate of development on temperature, (c) the effects of fruit age and temperature on developmental times, (d) the effects of temperature on survival, (e) the effects of relative humidity on survival, and (f) the effects of relative humidity on developmental time. [See Isley 1932.]*

$$R(\tau(t)) = a + b\tau \qquad [1a]$$

where a and b are regression constants fitted to the developmental rate data. Solving for the temperature where the rate of development is zero yields the lower thermal threshold ($\theta = a/b$). In theory, if a poikilothermic organism experiences different temperatures over the T days of its development, the rates must sum to unity over that period.

$$R_1(\tau_1) + R_2(\tau_2) + \cdots + R_T(\tau_T) = 1 \qquad [1b]$$

If temperature is changing rapidly, we could sum these rates over smaller intervals of time Δt in the same manner.

$$\sum_{t=1}^{\infty} R(\tau(t))\Delta t \approx \int_{t=0}^{\infty} R(\tau(t))\, dt = 1 \qquad [1c]$$

The Degree Day Rule The temperature summation rule [i.e., the degree day rule $(D°)$] was proposed in principle by Candolle (1855). This rule is easy to develop and use, and works well for many applications. This model is formulated as

$$D° = T(\tau)(\tau - \theta) \tag{2}$$

where $D°$ is the physiological time (degree days) required to complete development of a stage, T is days required to complete development at temperature τ, and θ is the temperature threshold for development estimated from [1a]. The confidence interval for θ may be computed using the procedures outlined in Campbell et al. (1974). Rearranging [2] gives us the equivalent developmental rate models.

$$R(\tau(t)) = \frac{1}{T(\tau)} = \frac{1}{D°}[\tau(t) - \theta] = b\tau(t) - a \tag{3}$$

This is the simplest and most widely used method for predicting physiological age and time for populations of poikilotherm organisms. In words, $D°$ is the product of time and the degrees of temperature above the threshold temperature. The computations for $D°$ under constant conditions are illustrated in Figure 3.3a (i.e., the cross-hatched area) for two temperatures ($\theta < \tau_1 < \tau_2$). Note that $D°_1 = D°_2$ are the same for both temperatures.

The theory is easily extended to fluctuating temperatures as indicated in Figure 3.3b, which shows daily patterns of temperatures and the daily integral of $\Delta D°(t)$ above θ, which are summed to determine when the constant $D°$ has been reached. One method to compute $\Delta D(t)$ is to force a sine curve through the maximum and minimum temperatures and integrate the area under the curve above θ (see Campbell et al. 1976; Allen 1976). A poikilothermic organism would be expected to complete its development when the sum

$$D° = \sum_{t=1}^{T} \Delta D°(t) \tag{4}$$

In practice, the linear model is quite adequate for most situations, but gross errors accrue in predicting development at extreme temperatures marginally favorable for the development of the organism.

Nonlinear Models

Ideally, we should know the response of the organism over the entire range of temperature (see Fig. 3.2a,b), and when available, we can accurately compute developmental rates for any temperature pattern. The oldest concept is van't Hoff's law (i.e., the Q_{10} rule), which states that for every 10°C increase in

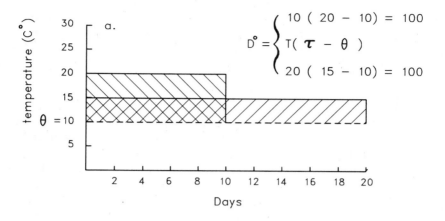

$$D^\circ = \begin{cases} 10\,(\,20\,-\,10) = 100 \\ T(\,\tau\,-\,\theta\,) \\ 20\,(\,15\,-\,10) = 100 \end{cases}$$

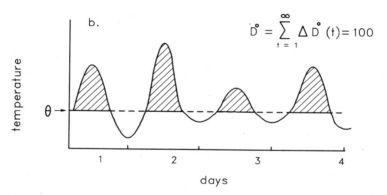

$$\dot{D}^\circ = \sum_{t=1}^{\infty} \Delta \dot{D}^\circ (t) = 100$$

Figure 3.3. *Illustration of the concept of degree days: (a) two constant temperatures (i.e., $\tau_1 = 15$ and $\tau_2 = 20°C$, and (b) variable temperatures. Note that degree days accumulate above the temperature threshold value of θ.*

temperature, there is a doubling or tripling of the rate of chemical reaction (van't Hoff 1884). In fact, the rates of chemical reaction may be lower or higher than predicted by Q_{10}. The Arrhenius (1889) equation derived from van't Hoff's law is written as

$$K_1 = K_o \exp[a(\tau_1 - \tau_o)/\tau_1\tau_o] \qquad [5]$$

where K_1 and K_0 are the rates of chemical reaction at absolute temperatures τ_1 and τ_0, respectively, exp is the base of the natural logarithm, and a is a constant bringing Q_{10} closer to 2.

Other empirical models could also be used. For example, Huffaker (1944) used the inverted from of Janisch's (1932) catenary model [6] to describe the developmental rates of *Anopheles* mosquito on temperature.

$$R(t) = \frac{2c}{a^{\tau - \tau_m} + b^{\tau_m - \tau}} \qquad [6]$$

where a, b, and c are empirical constants, τ is temperature, and τ_m is the temperature where the developmental rate is maximum. The Janisch model is the addition of two exponential functions, which as Huffaker explains . . . *means biologically that two things working opposingly are added, a positive function, the acceleration of developmental processes through rising temperatures, and a negative function, the retardation and injury in the process of development.*

Stinner et al. (1974) fit a sigmoid function to spotted alfalfa aphid developmental rate data (Messenger and Force 1963).

$$R[\tau(t)] = \frac{C}{1 + \exp[k_1 + k_2\tau(t)]} \qquad [7]$$

where τ is the temperature at time t

$$\tau = \begin{cases} \tau & \text{for } \tau < \tau^* \\ 2\tau^* - \tau & \text{for } \tau \geq \tau^* \end{cases}$$

and C is the maximum developmental rate at temperature τ^*, and k_1 and k_2 are constants. Above τ^* the rate of development slows, necessitating discontinuous values for τ. We should note that the maximum rate of development is not the same as the optimal one (see Chapter 4). Other models have been proposed that describe this relationship better (Larsen and Thomsen 1940; Stinner et al. 1975; Logan et al. 1976), but they are variants of the same idea. Most of the above models are, however, empirical and not based solidly in theory. Attempts to make direct analogies to chemical reaction rates must invoke theory that assumes the constants have thermodynamic significance.

A Biophysical Model

The biophysical model for poikilotherm development proposed by Sharpe and DeMichele (1977) and Schoolfield et al. (1981) is a radical departure from those based on empirical fits to data. In this model, thermal development is seen as a series of complex enzymatic reactions within the organism that natural selection has assured are well suited to its thermal environment. Their arguments are biochemical and are summarized as follows.

Consider that control points regulating the reaction rates are found either at the beginning or at branch points of the metabolic pathway. Enzymes located at these points are called control enzymes and regulate the rate of the overall metabolic process. They further assume that for convergent parallel pathways, the coordination will be such as to ensure sufficient metabolism. The model assumes:

1. A control enzyme determines the developmental rate of an organism.
2. The developmental rate is proportional to the product of the concentration of the active enzyme and the temperature-dependent rate constant providing a mechanism for temperature inhibition of development.
3. The control enzyme can exist at low (P_1) and at high (P_3) catalytically inactive states as well as a catalytically active state (P_2).

This model of enzyme activity is depicted in Figure 3.4 where the transition rates (k_i, $i = 1, \ldots , 4$) between states are assumed normally distributed with the mean transition rate of i described by the Erying equation [8].

$$k_i = \frac{K\,A}{h}\,\exp\left[\frac{\Delta S_i(\Delta H_i/A)}{R}\right]$$ [8]

where K is Boltzmann's constant, A, is absolute temperature, h is Planck's constant, ΔS_i is entropy of activation, ΔH is enthalpy of activation, and R is the gas constant. The probability of transfer from one state to another is exponentially distributed over a small interval of time $\Delta t = t + \Delta t - t$.

The equation describing the developmental rate [$R_D(A)$] under nonlimiting substrate conditions (i.e., optimal food) at absolute temperature A is

$$R_D(A) = \left\{ \frac{A\,\exp\left[\dfrac{(\phi - \Delta H_A/A)}{R}\right]}{1 + \exp\left[\dfrac{(\Delta S_L - \Delta H_L/A)}{R}\right] + \exp\left[\dfrac{(\Delta S_H - \Delta H_H/A)}{R}\right]} \right\}$$ [9]

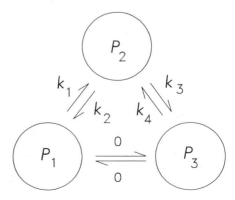

Figure 3.4. Enzyme activity states: P_2 is the active state and P_1 and P_3 are the low- and high-temperature inactivation states, respectively. The transition rates between the states are k_1, k_2, k_3, and k_4. Note that no transition occurs between states P_1 and P_3 except via P_2 (redrawn from Sharpe et al. 1977).

The exponential terms in the denominator account for the low- and high-temperature inactivation effects. The model has only one variable, the absolute temperature A and constants ϕ, ΔH_A, ΔS_L, ΔS_H, ΔH_L, and ΔH_H, which reflect the thermodynamic characteristics of the control enzyme system.

Unfortunately, these constants cannot currently be measured by biochemical methods, and as in the other models they must be estimated by least-squared methods from data. In this sense there is not much increase in the utility of this biophysical model over empirical ones, except that the biochemical/physical theory underlying this complicated phenomenon is eminently more satisfying.

THE EFFECTS OF NUTRITION

Sharpe and Hu (1980) introduced the effects of nutrition as a function of nitrogen-based nutrient substrate (i.e., protein or amino acids) into the model for poikilotherm development [10].

$$R_D(A, Q)$$
$$= \frac{Q}{Q + K_m} \left\{ \frac{A \exp\left[\frac{(\phi - \Delta H_A/A)}{R}\right]}{1 + \exp\left[\frac{(\Delta S_L - \Delta H_L/A)}{R}\right] + \exp\left[\frac{(\Delta S_H - \Delta H_H/A)}{R}\right]} \right\}$$

$$[10]$$

where K_m is the Michaelis constant ($= 5.67$) and Q is the average available nutrient intake rate. In simpler notation,

$$R_D(A, Q) = \frac{Q}{Q + K_m} \left\{ \frac{A \exp\left[\frac{(\phi - a_1/A)}{R}\right]}{1 + \exp\left[\frac{(a_2 - a_3/A)}{R}\right] + \exp\left[\frac{(a_4 - a_5/A)}{R}\right]} \right\} \quad [11]$$

with constants a_1, a_2, a_3, a_4, a_5, and ϕ usually estimated by least-squares regression. With the effects of nutrition added, [11] predicts well the developmental time when the whole life stage is spent on the same diet. The model does not accommodate time varying nutrition, and although the model advances theory, it is not a general solution to the problem.

Isley's (1932) experiments clearly show that cotton fruit age (i.e., nutrition), as well as temperature and relative humidity, have pronounced effects on boll weevil development (Fig. 3.2C,F) and survival (Fig. 3.2D,E).

Data on the nutritional effects of host fruit age on the rate of development of pink bollworm (*Pectinophora gossypiella* Saunders) (Fig. 3.5, Stone and Gutierrez (1986) support Isley's findings, and show the relationship between the age of bud (square) and large fruit (boll) attacked and pink bollworm developmental rate. The path of larval development is illustrated by the trajectories that begin at the left-hand edge of the figure (age of fruit) and reach the right-most edge at pupation. The fastest developmental rates have the shortest trajectories and occurs when large squares and medium sized bolls are attacked. An easy correction is to make the developmental rate a function of fruit age.

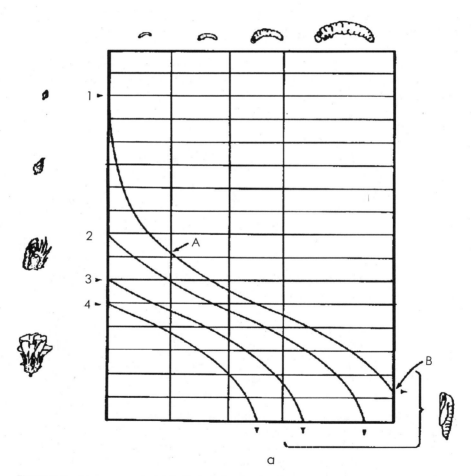

Figure 3.5. *A summary of pink bollworm development in different age fruit: (a) squares and (b) bolls. Fruit age is in the vertical axis and larvae age in the horizontal one. The paths follow the progress of individual larvae that at hatching enter fruits of different ages. Completion of development at pupation occurs when the paths reach the right-most edge of the figure. [Reprinted with permission from Stone and Gutierrez 1986.]*

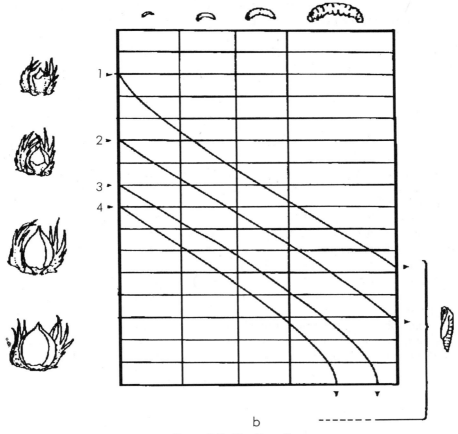

Figure 3.5. (Continued)

[As an aside, pink bollworm does not attack the vital parts of the square but rather feeds on the developing pollen making the abscission rate of infested and uninfested squares the same (Westphal et al. 1979). This enables the pest to use many of the infested squares 15 days later when they are postflower bolls.]

ESTIMATING THERMAL RELATIONSHIPS IN THE FIELD

Both linear (e.g., Nowierski et al. 1983) or nonlinear models have been used (e.g., Hochberg et al. 1986) to relate developmental rates of poikilotherms in the field to temperature, but the results have been checkered.

Suppose we had recorded the developmental times of individuals in n cohorts ($i = 1, \ldots, n$) under field conditions where temperatures varied within the linear range favorable for development. The linear model suggests that we would be able to estimate the threshold θ because the D_i° are in theory equal and time in days (T_i) depends linearly on τ_i as follows:

$$D_1^\circ = T_1(\tau_1 - \theta)$$

$$D_2^\circ = T_2(\tau_2 - \theta)$$

$$\vdots$$

$$D_n^\circ = T_n(\tau_n - \theta) \qquad [12]$$

One could estimate θ by solving the set of n equations using regression analysis or directly. This analysis presumes that the resource base is quantitatively and qualitatively constant and that temperatures remain in the favorable linear range. Fitting a nonlinear model is simply a bit more complicated. Nowierski et al. (1983) found for the walnut aphid [*Chromaphis juglandicola* (Kalt)] that the nutritional quality of walnut leaves varied over the season and this affected the developmental rate of the aphid even at the same temperature. The fit of the developmental rate model to the data might give an erroneous estimate of θ (see Fig. 3.6), but does such a finding question the whole concept of physiological time itself? The answer is unequivocally no, because despite this problem, physiological time provides a better time scale for examining poikilotherm development than does calendar time. In addition, the effects of nutrition can be included in the model (see Stone and Gutierrez 1986).

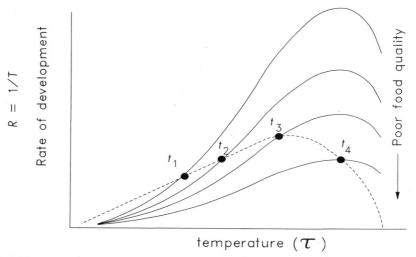

Figure 3.6. *Hypothetical rates (•) of development at different temperatures (τ) as modified by levels of nutrition (i.e., the solid lines). The dashed line is the function that might be extrapolated from such data.*

STOCHASTIC DEVELOPMENT

So far, we have dealt only with average developmental rates, in spite of the fact that biologists know that there are obvious differences among members of a cohort. These differences may be due to genetic, nutritional, microenvironmental, and other factors. Most population models use the average developmental rate, but several workers (Manetsch 1976; Stinner et al. 1975; Sharpe et al. 1977; Gutierrez et al. 1984; Plant and Wilson 1986; Curry and Feldman 1987) have proposed methods to incorporate distributed maturation rates in population dynamics models. Most of these methods simulate stochastic development (mean and variance), but they are in fact deterministic. A recent analysis of the Manetsch model by Severini et al. (1990) showed that it has a theoretical basis and can be tested against cohort data, but it is not explanatory of physiological or genetical mechanisms. The Manetsch model with modifications has proven quite useful for modeling field population dynamics and will be reviewed in this context in chapters 9–10.

MODELING DIAPAUSE DEVELOPMENT

The definition of diapause (dormancy or quiescence) in poikilotherm organisms is not entirely clear and readers are referred to Tauber et al. (1986) for a complete discussion of this topic. Here we use the term "dormancy" for this phenomenon in both plants and animals. What is clear is that there is more than one phase of development during dormancy, and each has a different relationship to temperature and likely other factors, such as drought, which may induce dormancy in tropical and arid climate species. The discussion here is restricted to temperature effects in temperate regions.

The rudimental approach outlined in this chapter was first proposed by Logan et al. (1979) for corn earworm [*Heliocoverpa zea* (Boddie)], and was later adapted by Johnsen et al. (in press) in their study of termination of dormancy in the cabbage rootfly [*Delia radicum* (L.)]. Figure 3.7 depicts three hypothetical phases of development that a dormant individual in a temperate climate must complete from late summer, when they entered dormancy, to early spring, when they emerge. (There may be more or fewer phases, and in some cases the dormancy may extend over several years.) In our hypothetical example, Phase 1 is a physiological process completed at low temperatures (chilling), ensuring that individuals entering dormancy in late summer will not emerge prematurely. Chilling is known in many temperate poikilotherms to be a prerequisite for breaking winter dormancy. Phase 2 is posited to occur at intermediate temperatures and serves to synchronize the development of cohorts during late winter and early spring. Phase 3 is simply postdormancy development and is the same as that experienced by nondormant individuals. Of course, factors such as nutritional histories affect the proportion of the population of susceptible individuals that enter dormancy as well as the lengths of their

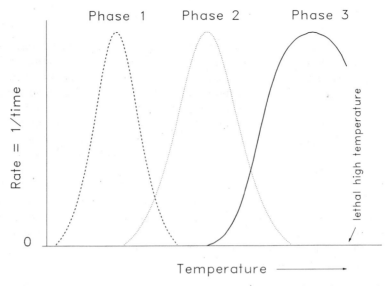

Figure 3.7. *Hypothetical rates of development during three phases of development of dormant poikilotherm organisms.*

dormant periods. Strong interaction of temperature, photoperiod, and nutrition on the rate of dormancy induction and termination has been documented in many organisms. Particularly good data sets from laboratory and field experiments exist for many organisms (e.g., pink bollworm, Adkisson et al. 1963; Gutierrez et al. 1981).

The important point is that a dormant individual must pass through all phases before it can complete its development, and hence the pattern of temperatures it experiences determines whether and how fast this occurs. In the aggregate, this development across all individuals determines the emergence pattern for the population. This model explains how premature emergence is prevented (Phase 1), how a population may attain a degree of synchrony (Phase 2), and how development proceeds when favorable conditions return (Phase 3).

EPILOGUE

The physical environment sets the ultimate limits for population growth. If a factor in the environment is limiting, a population may not develop (i.e., *Liebig's Law of the Minimum*). Temperature, especially in the organism's microenvironment, is an important factor that strongly affects the developmental rates of poikilotherms, but developmental rates may also be modified by nutrition, physiology, behavior, and genetics.

The use of temperature dependent time units for time, age, and all birth and death rates in modeling poikilotherm population dynamics is highly desirable (Hughes 1963), and is increasingly commonplace in the applied literature. Unfortunately, it has made little impact in the theoretical literature. The relationship of developmental rate to temperature is easily determined for most organisms, and where the complete relationship across temperature is available, a nonlinear model is desirable and essential at temperature extremes. In cases where the data are sparse, the simple degree-day approach outlined above usually yields satisfactory estimates in the linear range. Under special conditions, the relationship of developmental rate to temperature may be estimated from field studies, but caution should be used in accepting the results because they may be adversely affected by behavioral and nutritional factors. The theory of poikilotherm development based on models of enzyme kinetics appeals to intuition, but the direct application of the theory to field data is still remote and at present is a curve-fitting exercise.

4

Life Tables

After the flood, Noah placed adders on a log table and they began to multiply.
—Anonymous

Various types of life table analyses have been used to assess the effects of biotic and abiotic factors on population growth in nature (see Southwood 1975). Three types of life tables are apparent: laboratory age-specific life tables (see Andrewartha and Birch 1954) are used to estimate intrinsic birth and death rate parameters from a cohort of individuals under a given set of conditions, field age-specific life tables (Morris 1963; Varley and Gradwell 1960; Varley et al. 1973) estimate the effect of various factors in the field on population survivorship and reproduction, and field age-specific time-varying life tables (Hughes 1963, Gilbert et al. 1976; Manley 1989) provide time-varying estimates of population birth–death rates. Laboratory age-specific life tables are often used to estimate the effects of different levels of factors on vital rates (see below). A field age-specific life table consists of snapshots of birth–death rates under specific field conditions, and time-varying life tables are akin to a motion picture of the population birth–death rate dynamics, which are best executed as age-structured population dynamics models. These are important distinctions, and this chapter lays a historical and theoretical foundation.

Under nonlimiting conditions, a population (N_t) would experience exponential or Malthusian growth from some initial population of size N_0,

$$N_t = N_0 e^{r_m t}, \qquad [1]$$

where t is time and r_m is the population growth rate commonly called the Malthusian growth parameter. In nature, observed population dynamics are the result of effects of various biotic and abiotic factors that reduce birth and survivorship rates from the genetic (*intrinsic*) maximum to the observed (i.e., the transition of maximum age-specific vital rates to observed time-varying age-specific rates). Laboratory age-specific life tables are often used to estimate the intrinsic population growth rate parameter r_m, and the other life tables are used to estimate the observed final rate.

LABORATORY AGE-SPECIFIC LIFE TABLES

Discussions of laboratory age-specific life tables are found in many sources (Andrewartha and Birch 1954; Southwood 1975; Pielou 1969; Krebs 1978; Carey 1993), and are reviewed here for completeness. To construct a laboratory age-specific life table, cohorts of individuals are followed from birth under one or more sets of conditions and their age-specific survivorship and fecundity are recorded through the life cycle. A typical analysis consists of two tables: a survivorship budget and a life and fertility table.

A Survivorship Budget

The survivorship budget normally follows the fate of a large cohort of individuals, say 1000, and has the following components (Table 1):

l_x is the number alive at age x.
d_x is the number dying during the interval x to $x + 1$.
L_x is the number of animal between age x and $x + 1$.
T_x is the number of animals beyond age x.
e_x is the expectation of life remaining for individuals of age x.
$100q_x = 100d_x/l_x$ is the percentage dying during x.

Deevey (1947) and Slobodkin (1961) propose four types of survivorship curves (Fig. 4.1). In type I mortality occurs mostly on the old, in type II a constant number die per unit time, in type III the proportion dying per unit time is constant, and in type IV most of the mortality occurs in the early life stages. Type I is common for humans and type IV is typical for field populations of insects. As an aside, pesticide induced outbreaks of insect and arthropod pests

TABLE 1. A Hypothetical Survivorship Table

x (days)	l_x	d_x	L_x	T_x	e_x	$100q_x$
1	1000	250	875	2964	2.96	25.0
2	750	150	675	2089	2.78	20.0
3	600	125	537	1414	2.35	20.8
4	475	135	407	877	1.85	28.4
5	340	120	280	470	1.38	35.3
6	220	150	145	190	0.86	68.1
7	70	60	40	45	0.64	75.7
8	10	10	5	5	0.50	100.0

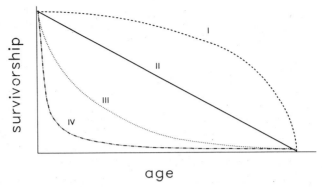

Figure 4.1. *Survivorship functions*

occur when their normal type IV survivorship pattern is disrupted due to the destruction of their natural enemies and is converted to a type I pattern.

An interesting statistic we can compute from these data is the expectation of life (e_x) beyond age x. If the age intervals are small, the number of animals (L_x) between ages x and $x + 1$ equals

$$L_x = \int_x^{x+1} l_x \, dx \qquad [2]$$

or approximately

$$L_x \approx \frac{l_x + l_{x+1}}{2} \qquad [3]$$

Summing the L_x beyond age x, we get

$$T_x = L_x + L_{x+1} + L_{x+2} + \cdots + L_\omega \qquad [4]$$

where ω is the maximum age. The expectation of life beyond age x is computed as

$$e_x = \frac{T_x}{l_x} \qquad [5]$$

In a hypothetical insect example (Fig. 4.2), the highest value of e_x is observed for pupae.

Life and Fertility Tables

Life and fertility tables (e.g., Table 2) are of greater interest in the context of this book as they allow us to estimate, among other statistics, the net repro-

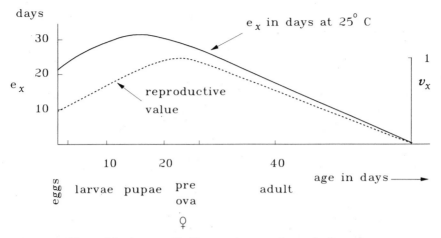

Figure 4.2. Age-specific life expectancy and reproductive value.

ductive rate per generation (R_0, female per female), the intrinsic rate of population increase (i.e., r_m), and Fisher's (1930) reproductive value v_x, which is a relative measure of the contribution of the different ages to the ancestry of future generations. The proportion of individuals surviving to age x is ℓ_x, and m_x is the age-specific number of females produced per female during the interval $(x, x + 1)$. We may compute R_0 (≈ 4.5) by substituting the ℓ_x, and m_x data in the approximate or exact formulas [6].

$$R_0 = \sum_{x=1}^{\omega} \ell_x m_x \text{ is the approximate method} \qquad [6a]$$

$$R_0 = \int_0^{\omega} \ell_x m_x \, dx \text{ is the exact method} \qquad [6b]$$

TABLE 2. Hypothetical Life and Fertility Table

Stage	Age (x = days)	Proportion Surviving (ℓ_x)	Females/Female (m_x)
Egg	0.5	1.00	0
Larva	3.5	0.92	0
Pupae	12.5	0.65	0
Adults	18.5	0.60	0
Adults	19.5	0.56	2.3
Adults	20.5	0.51	12.7
Adults	21.5	0.43	23.8
Adults	22.5	0.12	13.5
Adults	23.5	0.05	4.5
Adults	24.5	0	0

There are approximate and exact methods for computing r_m. Lotka's exact method involves solving [7] for r_m:

$$\int_0^\omega \ell_x m_x e^{-r_m x}\, dx = 1 \qquad [7]$$

Using the ℓ_x and m_x data in Table 2, r_m ($= 0.128$) was estimated using an error minimizing algorithm (e.g., Powell 1964). To do this requires an initial estimate for r_m and a predetermined criterion for goodness of fit. After the first calculation, the program determines whether the observed summation is sufficiently close to unity. If the deviation exceeds the criterion, the program then determines whether the initial r_m value should be incremented or decremented to improve convergence, and the calculations run again. This process continues until the result is sufficiently close to 1.

Because of its importance, a brief derivation of [7] is given to show origins (see Pielou 1969 for complete details). Lotka defined the average birth rate (b) of a population of N_t females at time t whose total births equal B_t as

$$b = \frac{B_t}{N_t} \qquad [8a]$$

He defined the proportion of N_t in age class x to $x + dx$ at time t as $c_x\, dx$, and hence the number of individuals in this age range is $N_t c_x\, dx$.

If the total birth during x to $x + dx$ is $B_x\, dx$ at time t, then $N_t c_x\, dx$ are the survivors of those born at time $t - x$. Specifically, the number alive today (N_t) of age x can be related to the number born x time units ago (i.e., $B_{t-x} = bN_{t-x}$) by the proportion surviving (ℓ_x):

$$N_t c_x\, dx = \ell_x B_{t-x}\, dx \qquad [8b]$$

Substituting terms and rearranging [8a] implies

$$B_{t-x} = bN_{t-x} = bN_t e^{-rx} \qquad [8c]$$

where e^{-rx} is the survivorship rate over the time interval t to $t - x$. Hence, if we substitute [8c] for B_{t-x} in [8b] and solve for c_x, we get

$$c_x = \ell_x b e^{-rx} \qquad [8d]$$

and of course

$$\int_0^\mu c_x\, dx = 1 \qquad [8e]$$

Substituting [8d] for c_x in [8e] yields

$$\int_0^\omega \ell_x b e^{-rx}\, dx = 1 \qquad [8f]$$

We now write the expanded definition of B_t where the average age-specific fecundity of females (m_x) of age x replaces the average value b:

$$B_t = \int_0^\omega N_t c_x m_x\, dx \qquad [8g]$$

Further substituting the definition [8d] for c_x in [8g] yields

$$B_t = \int_0^\omega N_t m_x \ell_x b e^{-rx}\, dx \qquad [8h]$$

and because b and N_t are constants, they may be taken outside of the integral. This yields Lotka's formula:

$$\frac{b}{b} = \frac{B_t}{bN_t} = \int_0^\omega m_x \ell_x e^{-rx}\, dx = 1 \qquad [8i]$$

With our hard won estimate of $r = r_m$, we can now estimate Fisher's (1930) reproductive value v_x. This parameter is a relative measure of the contribution of the different ages to the ancestry of future generations.

$$v_x = \frac{1}{\ell_x e^{-rx}} \int_0^\omega e^{-rx} \ell_x m_x\, dx \qquad [9]$$

In our hypothetical example (Fig. 4.2), new adults contribute the most to the ancestry of future populations.

Mean and Variance of Life Table Statistics

We might wish to determine the robustness of life table statistics (r_m, R_o, etc.) derived from a specific cohort. Estimates of the mean and variance of these statistics may be obtained using subsampling techniques (Efron 1982; Meyer et al. 1986). In this method, cohort members from the life table study are selected at random (*Bootstrap*) or systematically (*Jack-knife*), and the subsample used to compute a new set of life table statistics. Subsampling with replacement is done several times, and the set of life table statistics used to estimate means and variances as required (see Wermelinger et al. 1991b).

Practical Applications

In addition to estimating intrinsic vital statistics for use in demographic studies, a very appropriate use of such life tables is to assess the effects of harvesting of laboratory populations of natural enemies mass cultured for release and/or in the production of animals for food. Assessments of optimal rearing conditions are appropriate questions. For example, Chi (1988) developed an age-specific life table that included age-stage and two sexes, and Chi and Getz (1988) used it to evaluate harvesting strategies of laboratory populations.

There have also been many attempts to use laboratory age specific life tables to predict aspects of field population dynamics. Among these is the Mediterranean fruit fly (*Ceratitis capitata* Weidemann) eradication program in California, where Carey (1982) attempted to predict theoretically the number of generations required to eradicate the pest using pesticides. The analysis had little application to the field problem, but it did demonstrate the development of stable age distributions in populations growing under assumed nonlimiting conditions (Fig. 4.3).

Among the more successful applications of laboratory age-specific life table analyses to field problems were those of Messenger (1964, 1968), Messenger

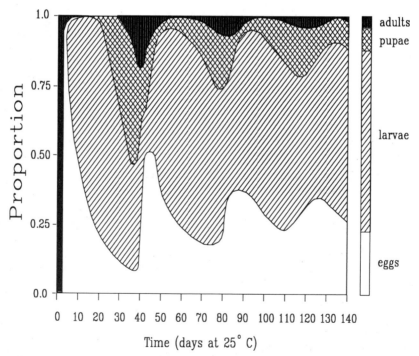

Figure 4.3. *The age structure dynamics of the Mediterranean fruit fly leading to a stable age distribution after day 140. [Redrawn with permission from Carey 1982.]*

and Force (1963), and Force and Messenger (1964). In those studies, several observed weather patterns were simulated in bioclimatic chambers and the life table statistics of three introduced parasitoids used to evaluate their potential for the biological control of the then newly introduced spotted alfalfa aphid (*Therioaphis maculata* Buckton). (The control of this aphid is thought to have occurred via the combined action of host plant resistance and introduced natural enemy activity.) Some of Messenger's results shown in Figure 4.4 compare the net reproductive rates (R_o) and intrinsic rates of increase (r_m) of the species at different temperatures (Force and Messenger 1964). The maximum R_o for all three parasitoids occurs near the same temperature, but the magnitudes and ranges of favorable temperatures are quite different. The maximum r_m for the aphid occurs at temperatures higher than those of the parasitoids (Fig. 4.4c): *T. maculata* > *Trioxys utilis* > *Praon palitans* > *Aphelinis semiflavis*.

In California, the aphelinid parasitoid *A. semiflavis* is restricted to the coastal and near coastal regions where the weather is cooler and frosts are unlikely; the distribution of the aphidiid *P. pallitan* extends into the near coastal valleys, while *T. utilis* is ubiquitous throughout the aphid's range including the great central valley where temperatures are very high in summer and heavy frosts occur in winter (van den Bosch et al. 1964; Gutierrez and van den Bosch 1971).

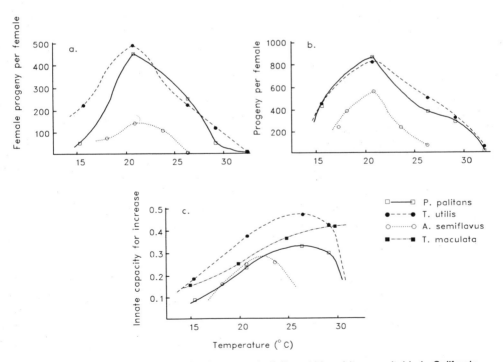

Figure 4.4. Life table statistics for the spotted alfalfa aphid and its parasitoids in California alfalfa.]Reprinted with permission from Messenger and Force 1963 and Force and Messenger 1964].

The wider degree of overlap of *T. utilis* parameters with those of its host explains in part why it is the most common and "probably" the most effective of the parasitoids throughout the aphid's range.

But such statistics may prove misleading, because other aspects of the natural enemies' biology may preclude their effectiveness in the field. For example, *Trioxys pallidus* (Haliday) was introduced from France to California for the control of walnut aphid (*Chromaphis juglandicola* Kalt) (van den Bosch et al. 1970), but it failed miserably throughout the hot dry regions where most commercial walnut is grown. However, a related biotype that could aestivate during the hot dry summer was introduced from Iran and it quickly brought the aphid under extremely good biological control in all areas (van den Bosch et al. 1970).

Similarly, Flint (1981) examined the life table statistics of the two biotypes of *Trioxys utilis*, a parasitoid of the spotted alfalfa aphid, and discovered that they were essentially the same. Again, an Iranian biotype had the capacity to enter aestival diapause, and this proved to be the essential component in its biology allowing it to survive the hot dry summers in California when aphid hosts are rare. No doubt other more subtle physiological and behavioral differences occur between the two biotypes, but these could not be discerned using life tables. Figure 4.5 compares the climate at Fresno (California), Teheran (Iran) and Rome (Italy), which shows the striking climatic match, and

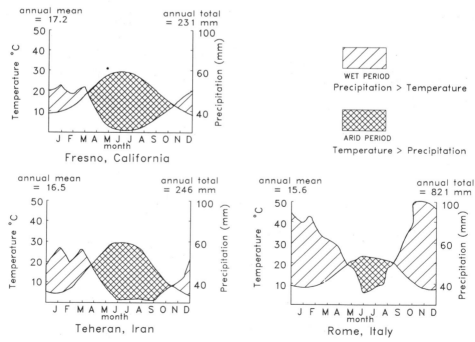

Figure 4.5. *Comparison of weather patterns from Fresno (California), Rome (Italy), and Teheran (Iran). [Reprinted with permission from Flint 1981].*

appear to explain why the Iranian biotype was preadapted to conditions in the Central Valley of California. Weather (and climatic) effects on organisms cannot be overlooked as they may often have a major impact on their dynamics and, in the extreme, determine the geographic range (i.e., the fundamental niche).

The searching and feeding behavior of organisms also may not be reflected in laboratory life table statistics and yet may have important consequences to field population dynamics. The life table statistics of the cassava mealybug parasitoid *Epidinocarsis lopezi* were judged to be inadequate for this parasitoid to control its host in West Africa (Odebiyi and Bokonon-Ganta 1986); a result clearly at odds with the facts. Control of the mealy bug was due to the combined effects of host size preferences, the parasitoid's host feeding behavior, changing plant nutrition and stochastic weather effects on the host, and in addition, the fact that *E. lopezi* found hosts at very low densities (Hammond et al. 1987; Gutierrez et al. 1988b, 1994) by cueing in on host kairomones (e.g., van den Meiracker et al. 1988). Obviously, these factors cannot be included directly into age-specific life table analyses, but must be considered if field population dynamics are to be explained. This requires another kind of life table analysis.

The Effects of Temperature and Nutrition on Life Table Statistics

The microclimate (e.g., temperature and humidity) experienced by a small organism may differ radically from ambient and also have profound influences on life table statistics (Baumgärtner and Severini 1987). Temperature affects not only developmental rates (see Campbell et al. 1974; Curry and Feldman 1987), but also age-specific fecundity (m_x) and survivorship (ℓ_x) and hence r_m. This can be seen in the extensive laboratory age-specific life table data on the blue alfalfa aphid, *Acyrthosiphon kondoi* Shinji (Fig. 4.6a–c, Summers et al. 1984). The r_m values computed on a per day basis are concave with respect to temperature (Fig. 4.7). However, the rates converted to a physiological time scale are the same over the middle range of favorable temperatures, and decline only at the extremes. The results suggest that on the aphid's time scale, the effects of temperature on within generation population growth rates are not important, but there may be adverse carry-over effects to the next generation as total fecundity and size and average survival times in the poikilotherms may have parabolic relationships to temperature (Fig. 4.6d–e). The effects of temperatures in one generation may be expressed in the next generation.

In a field example, MacKay et al. (1989) found that body length in adult pea aphids collected during the hot Australian summer was as much as 30% less than in fall and spring populations (Fig. 4.8). Reductions in body length of this magnitude translate to 50–60% reductions in body mass and fecundity. Thus, seasonal fluctuations of temperatures and other factors that affect size may have important demographic consequences for poikilotherms as undersized adults are likely to produce smaller and fewer offspring (Dixon 1987). In general, size and fecundity are indices of environmental favorability and are

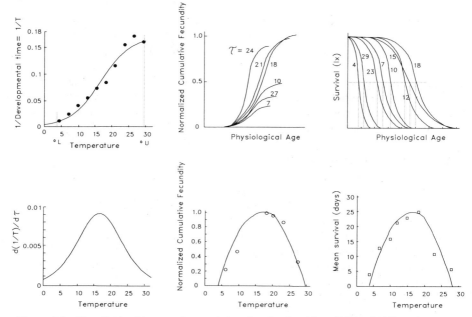

Figure 4.6. The effects of temperature and abundant food on blue alfalfa aphid (Acyrthosiphon kondoi Shinji): (a) developmental rates, (b) patterns of cumulative reproduction, (c) patterns of survivorship, (d) acceleration of developmental rates, (e) cumulative fecundity, and (f) time to 50% mortality. [Reprinted with permission from Summers et al. 1984.]

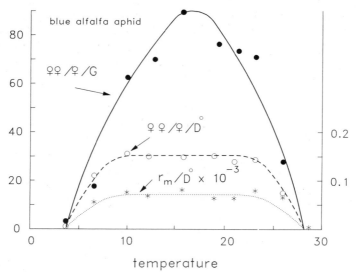

Figure 4.7. Life table statistics for blue alfalfa aphid at different temperatures. Note that R_o is plotted per generation (G) and per physiological age (degree days, $D°$, see also Fig. 4.6).

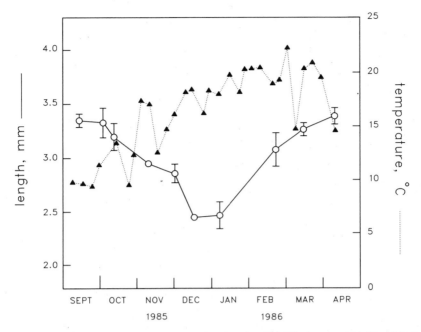

Figure 4.8. *The effect of field temperatures on the size (length) of pea aphid. [Reprinted with permission from MacKay et al. 1989].*

not reflected in life table studies not designed to investigate between generation effects.

The nutrient status of hosts may also have important demographic consequences. For example, Wermelinger et al. (1991b) showed in the two-spotted spider mite that r_m was correlated to concentrations of N in leaves. Fourfold increases in r_m and 11-fold increases in R_o occurred over the range of 1.5–3.0% leaf nitrogen. The importance of nutrition on the field dynamics of mites has been demonstrated in field experiments (Yaninek 1985; Wilson et al. 1988; Yaninek et al. 1989).

FIELD AGE-SPECIFIC LIFE TABLES OF SPECIES WITH NONOVERLAPPING GENERATIONS

The construction of life tables in the field is more difficult, especially for species with overlapping generations. In this section, we review the development of field life tables for species that ideally have the following properties: one generation per year, the life stages are easily separable, the sources of mortality are readily identifiable, and their impact accurately measurable (Varley et al.

TABLE 3. Abbreviated Mean Life Table for Spruce Budworm at Green River New Brunswick, Canada[a]

(x) = Age(x)	(lx) = Number Alive at x	Factor Responsible	(dx) = Number Dying During x	dx = % (100 qx)
Egg	200	Parasites	10	5
		Other	20	10
		Total	30	15
Early larva	170	Dispersal	136	80
Later larvae	34	Parasites	13.6	40
		Disease	6.8	20
		Other	10.2	30
		Total	30.6	90
Pupa	3.4	Parasites	0.3+	10
		Other	0.5+	15
		Total	0.8	25+
Adult moth	2.5	Miscellaneous	0.5	20

[a]Modified from Morris, 1963
[b]Generation survival 2 (1%), generation mortality 198 (99%), and sex ratio taken as 1 : 1 (males : females).

1973). Table 3 is an example of a field life table developed by Morris (1963) for the spruce budworm in Canada. Such field life table data have been vital elements for assessing the role of biotic and abiotic factors on the birth and death rates of such species.

Varley et al. (1973) and Morris (1963) developed *k-value* and *key factor* methods, respectively, to analyze field life table data. However, Luck (1971) concluded that the Morris technique could not be used to detect the presence of density dependence. This is important because only mortality factors that operate in a density dependent manner can regulate a population. For this reason, only the Varley et al. (1973) method is reviewed here. Interested readers should consult Manley (1989) for a complete discussion of these methods.

The Concept of *k*-Value

The concept of *k*-value ("killing power," Fisher 1930) is defined by Varley et al. (1973) as *the logarithm[1] of the ratio of the initial number (N) at the time the mortality factor began its action to the number of survivors (N$_s$).*

$$k = \log_{10} \left(\frac{N}{N_s} \right) = \log_{10} N - \log_{10} N_s \qquad [10]$$

[1]In review, the log of a number is the power the base must be raised to obtain that number (e.g., $10^0 = 1$, $10^1 = 10$, $10^2 = 100$, and $10^3 = 1000$). Of course, one can use the base e, then $e^{6.0977} = 1000$, or any other base). Obviously, the base 10 yields the most intuitive results.

The meaning of k-value, however, is not readily apparent from the definition, hence it is reviewed by example.

Suppose for simplicity, we formulate a simple *difference equation* model for a population where only one mortality factor is operative. If the number in generation $g + 1$ is $N(g + 1)$, F is the per capita reproductive rate of parent individuals $N(g)$ in the current generation and $S(g)$ is their survivorship rate, our simple population dynamics model may be written as follows:

$$N(g + 1) = FN(g)\, S(g) \qquad [11]$$

Taking the logarithms of both sides yields

$$\log_{10} N(g + 1) = \log_{10} F + \log_{10} N(g) + \log_{10} S(g) \qquad [12]$$

The link of k-values to this model is via our survivorship term $S(g)$. It is easy to visualize what the concept of k-value means by examining a specific functional response model such as [13a] (Nicholson and Bailey 1935; see Chapter 5) which estimates the mortality caused by parasitoids. In this model, N_a is the number of hosts (N) attacked, P is the number of parasitoids, and s is the per capita search rate.

$$N_a = N(1 - e^{-sP}) \qquad [13a]$$

Rearranging and defining N_s as survivors yields

$$e^{-sP} = \frac{N - N_a}{N} = \frac{N_s}{N} \qquad [13b]$$

e^{-sP} is a modified form of the zero term of a Poisson distribution or the probability of escaping attack from P parasitoids (our survivorship term, S). If [13b] is converted to base 10 using a constant (2.3)

$$\frac{N_s}{N} = 10^{-(sP/2.3)} \qquad [13c]$$

and rearranging and taking the logarithm yields our definition of k-value for a specific value of P.

$$k_p = \frac{sP}{2.3} = \log_{10}\left(\frac{N}{N_s}\right) \qquad [13d]$$

The negative of the exponent of our survivorship is our k-value and it may be a function (say $k_P = sP$). Hence, $0 < S(g) = 10^{-k_P(g)} < 1$ is the proportion of $N(g)$ surviving, and by substituting its logarithm in [12] we get

$$\log_{10} N(g + 1) = \log_{10} F + \log_{10} N(g) - k_P(g) \qquad [14a]$$

If we had two (or more) factors, say one being density dependent (k_{dd}) and the other density independent (k_{di}), the resulting model would be

$$\log_{10} N(g + 1) = \log_{10} F + \log_{10} N(g) - k_{dd}(g) - k_{di}(g) \qquad [14b]$$

Taking the antilog of [14b] gives

$$N(g + 1) = F \cdot N(g)10^{-k_{dd}(g)}10^{-k_{di}(g)} = F \cdot N(g)S_{dd} \cdot S_{di} \qquad [14c]$$

Of course survivorship rates are multiplied, but on a log scale they must be added (e.g., $K = k_1 + k_2 + \cdots + k_n$ for $j = 1, \ldots, n$). The important point is that writing the model in terms of logarithms and k-values is the same as writing it in the more familiar survivorship form. The utility of k-values is that one can estimate the type and magnitude of the mortality from field data.

k-VALUES AS FUNCTIONS

It would be nice if nature were predictable so that simple relationships could be found from life tables for predicting the different k-values (e.g., the Nicholson–Bailey model above). Suppose we had several years of life tables and we plotted the components k_j of total mortality K on time (see Fig. 4.9). In our example, the pattern of k_3 is most similar to that of K, suggesting that it is the major determinant of the overall population patterns. The other two k-values in our example fluctuate, but clearly not in the same way as K. However, for factor 3 to regulate, its intensity of action (its rate) must increase with the density of the population upon which it acts.

Estimating Density Dependence from Field Life Table Data

Varley et al. (1973) propose that plots of k-values on the log density on which the factor acts can be used to determine the type of interaction between a population and its mortality agent. Such data may yield a wide variety of relationships (Fig. 4.10). Relationships with significant positive slopes are assumed to act in a direct density dependent manner, those with negative slopes in an inverse density dependent manner, and those with slope zero are said to be density independent. If a relationship is described by a linear function of the form $k_j = a_j + b_j \log N$, the factor is overcompensating if the slope of the function (b_j) is greater than unity, perfect density dependent if it is unity, and under compensating if it is less than unity. Increasing values of b_j may cause

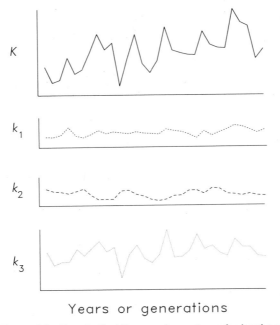

Figure 4.9. *Hypothetical time varying patterns for k-values.*

population fluctuations to increase (see Varley et al. 1973). Last, the intercept *a* could be interpreted as a threshold level for the mortality factor.

A population model (e.g., [14]) can be built with the *k*-values being functions of predator (P) and prey (N) density. As an exercise, assume that host fecundity $F = 100$, and that we have three mortality factors $(j = 1, 2, 3)$, with *k*-values $k_1 = 0.3P(g)$ (under compensating direct density dependence), $k_2 = 0.8 - 0.25\log N(g)$ (inverse density dependence), and $k_3(g)$ being some random variable (density independent, i.e., rainfall). The time projections of the host and the mortality caused by each mortality factor separately may be computed and is left as an exercise (see Varley et al. 1973).

Partitioning Mortality by Factor

Partitioning the mortality (N_a) among the j factors is not difficult. The Venn diagram in Figure 4.11 illustrates the problem and also explains the concept of dispensable and indispensable mortality. If the circle of unit 1 is the probability of survival during the period g to $g + 1$ in the absence of natural enemies P_1 and P_2, then the combined survival rate (S_T):

$$0 \leq S_T = S_1S_2 = [(1 - \mu_1)(1 - \mu_2)] = 1 - \mu_1 - \mu_2 + \mu_1\mu_2 < 1 \quad [15]$$

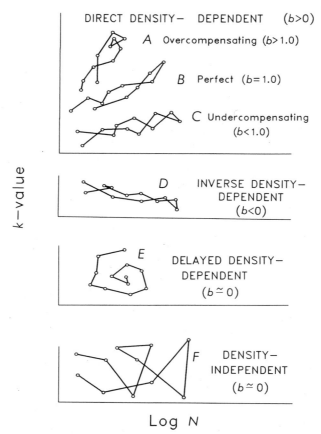

Figure 4.10. *Plots of hypothetical k-values on log$_{10}$ host (see text for discussion). [Reprinted with permission from van den Bosch et al. 1982.]*

where $0 \leq \mu_j \leq 1$ is the jth mortality rate. The proportion $\mu_j - \mu_1\mu_2$ is the indispensable mortality, or the mortality that would not occur if the other was absent, and $\mu_1\mu_2$ (i.e., the area of overlap) is the dispensable mortality, or that which would occur in the absence of one of the natural enemies. This assumes that the factors are acting simultaneously and that they do not discriminate between previously attacked hosts. If the factors discriminate perfectly, then for a sufficiently short-time period, the species get μ_1 and μ_2 proportion of the hosts, respectively, but $\mu_1 + \mu_2 < 1$. However, if P_1 always wins over P_2 in cases of multiple parasitism, then P_1 gets μ_1 of the hosts and P_2 gets $\mu_2 - \mu_1\mu_2$. If one species is a parasitoid and the other is a predator that also attacks parasitized hosts, then the computations become a bit more complicated, but not excessively so, and they are left for the reader as an exercise.

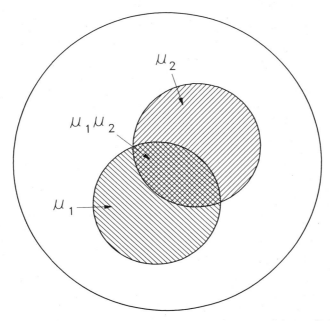

Figure 4.11. *Venn diagram of two interacting sources of mortality.*

Applications of Age-Specific Field Life Tables

A recent and comprehensive field age-specific life table study describing the dynamics of an insect species is Münster–Swendsen's (1985, 1991) work on the spruce needle miner, *Epinotia tedella* (Cl.) in a Danish forest (Fig. 4.12a). This pest was studied over a period of 20 years, and included records of adult and larval densities and of natural enemies (parasitoids and predators). Mortality due to parasitoids, predators, a fungal disease, and the effect of a sublethal protozoan infection were measured and their impacts analyzed with respect to densities of the causal agents and weather. As a new innovation in this kind of life table, the condition of the host tree was measured as a function of previous precipitation. In addition, postfeeding larval dry weights were recorded and included in the analyses. Unlike previous studies, no large sources of *unexplained disappearance* (mortality) were observed, and the data provided an excellent basis for the construction of a model to simulate the dynamics of the pest.

The herbivore dynamics are cyclic with delayed density dependent mortality due to parasitoids (Fig. 4.12b). A simple submodel for parasitism derived from the Nicholson–Bailey (1935) model simulated the observed dynamics perfectly with respect to the oscillation period. However, additional mortality, biological negative feedback mechanisms, and varying fertility were equally important.

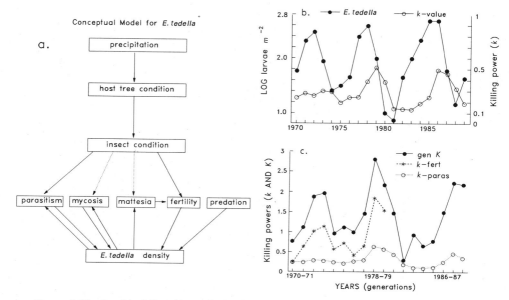

Figure 4.12. *Graphical life table analysis of the spruce needle miner: (a) [Reprinted with permission from Münster-Swendsen 1991.] The conceptual model of interacting factors, (b) Cyclic herbivor dynamics with delayed density dependent mortality due to parasitoids and (c) K and separate k-functions for parasitism and fertility. [Reprinted with permission from Münster-Swendsen 1985.]*

In fact, the reduction in fertility expressed as a k-function varies in parallel with parasitism, and was the prime factor (Fig. 4.12c). The inclusion of variable fertility as a function of population density provided a highly satisfactory description of both period and amplitude of the pest population oscillations.

Limitations of the *k*-Value Method

It is insufficient to simply estimate the amount of mortality itself, as the observed mortality rate may vary over time and be the result of interactions of, among other things, age structure, nutrition, behavior, and possibly lower and higher trophic levels as modified by abiotic factors (e.g., weather). The k-value analyses are useful in sorting out these factors, but their use is generally limited to populations with discrete generations and where the impact of the mortality factors can be measured accurately. Often the latter criterion is not met for many species, especially where the generations and life stages overlap and the resource base is changing rapidly in quantity and quality over time. The most important limitations are the excessive length of time and expense involved in such studies, and the results are time and place specific. Properly formulated, time varying life tables provide a suitable alternative, and hence the origins are reviewed for strictly historical reasons.

TIME-VARYING LIFE TABLES

Hughes (1963) tried to cope with some of the complexity of species with overlapping generations using methods adapted from human demography. In doing this he invented the notion of a *time-varying life table* wherein age-structure and physiological time for both time and age were included. These innovations were a radical departure from previous methods and despite criticisms (Carter et al. 1973) made significant contributions to the study of field population ecology.

Hughes assumed, and in many instances found, that rapidly growing populations of the cabbage aphid, *Brevicoryne brassicae* L., often achieved a stable age distribution. Under this assumption, the population would on a physiological time (t) scale exhibit exponential or Malthusian growth at the intrinsic rate of increase (λ).

$$N_t = N_o e^{\lambda t} \qquad [16]$$

If this occurs, then one progeny is produced at time $t_0 = 0$ and $e^{\lambda t}$ are produced at time t and the ratio of those aged 0 and t is $(1 : e^{\lambda t}) = e^{-\lambda t}$. Then, in general, the number of individuals aged t_1 to t_2 equals

$$\lambda N_t \int_{t_1}^{t_2} e^{-\lambda t}\, dt = \frac{1}{\lambda}\, [e^{-\lambda t_1} - e^{-\lambda t_2}]\lambda N_t = [e^{-\lambda t_1} - e^{-\lambda t_2}]N_t \qquad [17a]$$

Hence, if $t_1 = 0$ and $t_2 = 1$ defines the age interval for the youngest stage, their numbers equal

$$N_t[1 - e^{-\lambda}] \qquad [17b]$$

If the adult age class is defined by mean time to reproduction β and mean longevity ω, then their numbers are computed as

$$\lambda N_t \int_{\beta}^{\omega} e^{-\lambda t}\, dt = [e^{-\lambda \beta} - e^{-\lambda \omega}]N_t \qquad [18]$$

The per capita adult reproductive rate (γ) can be calculated as

$$\gamma = \frac{1 - e^{-\lambda}}{e^{-\lambda \beta} - e^{-\lambda \omega}} \qquad [19]$$

where $N_t(1 - e^{-\lambda})$ is the female progeny produced during one time unit and $N_t(e^{-\lambda \beta} - e^{-\lambda \omega})$ is the number of females in the reproductive age interval (β, ω). In nature, γ has two components: the maximum reproductive rate per female and selective mortalities that may operate on adults to reduce this value to the observed. Among the selective mortalities may be nutritional effects, diseases, parasitism or other factors that may decrease adult fecundity.

Computing Population Growth Rates from Field Data

Assuming a stable age distribution, the ratio of the numbers in the first three instars is a reasonable estimate for the finite rate of population increase (e^λ):

$$e^\lambda = \frac{N_1 + N_2}{N_2 + N_3} \qquad [20]$$

However, after the selective mortality rates operate, the observed rate of increase (e^ρ) at time t may be estimated graphically from field data plotted on a physiological time scale.

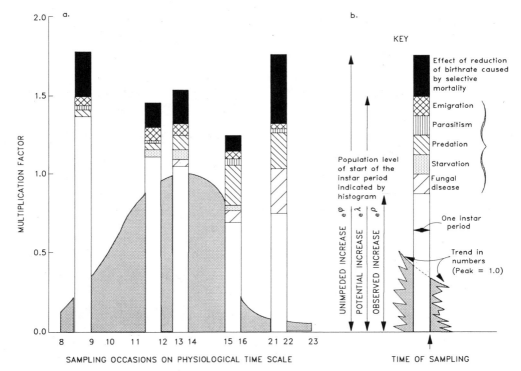

Figure 4.13. *Illustration of the components of a time varying life table: (a) time specific components of mortality affecting population growth rates during several time periods and (b) shown in detail for one sampling period. The dark area outlines the normalized trends in cabbage aphid numbers at different sampling times, the bars depict the maximum multiplication rate over one instar period (e^φ) and the sources of mortalities that reduce this rate. [See text, reprinted with permission from Hughes 1963.]*

$$e^\rho = \frac{N_t}{N_{t-1}} \qquad [21]$$

The difference $\mu = \lambda - \rho$ is the selective mortality rate due to all factors at physiological time t. Of course, during a sampling period, this mortality may come from many causes as illustrated in Figure 4.13b and the impact may be different at different times, hence similar computations need to be made for all sampling periods (Fig. 4.13a). Because λ and ρ are estimated from field data, another theoretical unimpeded rate φ is indicated in Figure 4.13b. The notion of a theoretically maximum growth rate is important and plays a role in the development of time varying life table model in subsequent chapters. The action of factors such as intraspecific competition for food, the inefficiency of search, subtle changes in food quality, and other biotic and abiotic factors may be difficult to assess with field samples. In addition, these small effects may propagate over time to introduce large errors in the model.

EPILOGUE

Life tables may be used to estimate population statistics under specific laboratory or field conditions, and yet what we wish for are methods that are independent of time and place and that are dynamic and yet general. Under optimal conditions, laboratory life table parameters may estimate the population's genetic potential rate of growth (Hughes' *unimpeded potential*), but this ideal is rarely met in nature. More likely, the estimate we get is the *potential rate* that in nature is further compounded by mortality factors to yield the *observed rate* (Gilbert et al. 1976).

The k-values derived from field life table data are useful only for assessing mortality in populations with non overlapping generations such as those common in north temperate regions. The method requires too much time and effort to gather the appropriate data. Hughes' method for estimating time varying demographic parameters for populations with overlapping generations has the necessary assumption of a stable age distribution, which greatly limits its application to field problems. This method is also difficult to implement and is time and place specific. His work, however, laid the basis for later attempts to develop more realistic time varying life table models (i.e., age structured population models), the first of which were developed by Hughes and Gilbert (1968) and Gilbert and Gutierrez (1973).

5

Resource Acquisition in Predator–Prey Systems

. . .the actual ratio of food eaten by a predator over a certain period of time will, under favorable feeding conditions approach a definite size, above which it cannot under any circumstances increase and which corresponds to the physiological condition of full saturation.
—Ivlev, 1955

In nature, all species are faced with evolving stable strategies for resource acquisition (i.e., the functional response, Solomon 1949) and allocation of resources to respiration, growth, and reproduction (i.e., numerical response). This biology can only be represented in a simplified form in population models. Any attempt at a one-to-one representation would make the model as complicated as nature itself. In developing a model, one should be cognizant of the trade-off between simplifications made for the sake of mathematical tractability and those made because the biology allows it. The mathematical underpinnings of many functional response models currently used in population ecology are reviewed by Royama (1971, 1992) and Hassell (1978). Here a simple population model is introduced to identify the functional and numerical response components.

For example, general first-order differential equations (i.e., Lotka–Volterra models) have been used to describe simple predator–prey dynamics [1]:

$$\frac{dN(t)}{dt} = f(N)N - g(N, P)P \qquad [1a]$$

$$\frac{dP(t)}{dt} = hg(N, P)P - \mu P \qquad [1b]$$

where $f(N)$ is the per capita birth rate of prey (N), $g(N, P)$ is the predator (P) per capita consumption rate of prey, h converts prey eaten to predator offspring, and μ is the per capita predator death rate (e.g., Arditi and Ginsburg 1989). Despite its simplicity, the model has heuristic value as it serves to introduce the concepts of per capita functional response [$f(N)$ and $g(N, P)$ for prey and predator, respectively] and predator numerical response [$hg(N, P)$]. The constant h is the number of predators produced per prey attacked, and equals unity in host–parasitoid interactions in which a single parasitoid emerges, but it may be greater than unity for certain parasitoid biologies (e.g., polyembryonic or gregarious species) and less than unity for true predators.

The functional response has been termed the prey death rate (Hassell et al. 1976) and the numerical response is the predator birth rate (Beddington et al. 1976), which includes predator aggregation to prey density via movement within or between systems. The functions f and g may be prey dependent [i.e., $g(N)$ or $g(P)$] or may depend on both N and P [i.e., $g(N, P)$] or their ratio [i.e., $g(^N/_P)$, *ratio dependent*] (Matson and Berryman 1992). Ratio dependence simply means that the number of prey killed per predator depends on the ratio of predator to prey (Berryman 1992). The notion of ratio dependence has been in the literature for decades (e.g., Thompson 1939; Leslie 1948; Frazer and Gilbert 1976; Gutierrez and Wang 1977; Gutierrez et al. 1981, 1984; Berryman 1981; Berryman and Stenseth 1984; Getz 1984; Arditi and Ginzburg 1989; Arditi et al. 1991; Berryman 1992), and only recently has it raised controversy when used in the context of differential equations (Abrams 1994).

THE FUNCTIONAL RESPONSE OF PREDATORS-TO-PREY DENSITY

In general, ecological theory supposes that predation (parasitism) success increases with host density, but there are exceptions to this rule, especially at high densities of prey when the rate of predation may decrease because of interference from excessive prey or the fouling of the habitat (e.g., excessive honeydew by Homoptera). Several functional response models have been proposed and some common ones are summarized in Table 1. All of these models have deficiencies that affect their utility (see Hassell et al. 1976; Gutierrez et al. 1976; Wang and Gutierrez 1980; Berryman et al. 1995). Very good reviews of this topic are found in Royama (1971) and Hassell (1978). Hence, only salient points are discussed here.

There are commonly thought to be three forms for $g(\cdot)$ (types I, II, III, see Fig. 5.1). Some type II and III models may also be ratio dependent. In general, the number of prey (hosts, N) attacked, eaten, or parasitized by a population of predators P is denoted by N_a and is given by the expressions

$$N_a = g(N)P \qquad \text{type I}$$
$$N_a = g(N, P)N \qquad \text{types II and III} \qquad [2]$$
$$N_a = g(P/N)N \text{ or } g(N/P)P \qquad \text{ratio dependent types II and III}$$

TABLE 1. A Partial Summary of Functional Response Models[a,b]

Author	Number Attacked	Type[c]	Contribution
Lotka–Volterra	$sPNT$	I(S)	First interaction model
Thompson	$(1 - e^{-sPT/N})N$	II(P,S)	Introduced probability theory
Nicholson–Bailey	$(1 - e^{-sPT})N$	I(P,S)	Introduced concept of search
Holling	$sPNT/(1 + sT_hN)$	II(P,S,B)	First to add behavior
Rogers (Para. i)	$(1 - e^{-s(P \cdot T_hNa)})N$	II(P,S,B)	Modified Holling equation to estimate the instantaneous probability of attrition
(Pred. ii)	$(1 - e^{-sP(T - T_hNa)})N$		
Hassell–Varley	$(1 - e^{-sTP^{1-m}})N$	I(P,S,B)	Incorporate parasitoid mutual interference
Frazer–Gilbert	$[1 - e^{-DPT(1 - e^{-SN(DP)/N})/N}]N$	II(P,S,B,D)	Added demand rate to model
Hassell	See Roger's models where $s = cN/(1 + dN)$	III(P,S,B,t)	Showed the transition from type II to type III
May	$[1 + (sP/K)^{-k}]N$	I(P,S,)	Added prey distribution to $f(N,P)$
Ivlev	$DP(1 - e^{-sN/D})$	II(P,S,N,D)	Added predator demand
Watt	$aP(1 - be^{-sNP^{1-\beta}})$	II(R,P,S,t)	Ratio dependent
Gutierrez-Baumgartner	$DP[1 - e^{-(sNP^{1-\beta}/D}]$ $\alpha = \alpha(t,\text{mass,age})$, $D = D(t,\text{mass,age})$,	II(R,P,S,D,M,t,D)	Added physiology and physiology-motivated behavior
Berryman	bP/N	II(R)	Simple ratio dependence

[a]See also Berryman et al. 1995 for ratio dependent models.

[b]N = number of prey, P = predators, s = search rate, T = total time, and T_h = handling time per prey, a, b and β are parameters.

[c]D = Demand rate; M = physiology; B = behavior; P = probability theory; R = ratio dependence, S = search theory; t = time-varying parameters, α = prey rate of apparency.

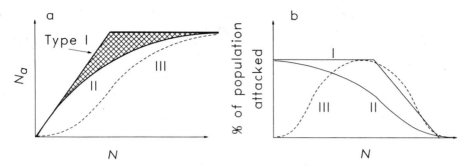

Figure 5.1. *Types of functional response models: (a) the per capita number of prey (N) killed (N$_a$) per predator (P), and (b) the proportion of the prey population killed by a predator as prey density increases. A maximum N$_a$ is assumed for the type I model (in Fig. 5.1a), and the crossed-hatched difference between types I and II is a measure of the search overlap among natural enemies or due to handling time.*

Though cumbersome, the notation N_a for number (or mass) of prey attacked is used here to conform with historical usage, although a has been used for area of search and for age. The term N_a may represent the quanta of light captured by plants and converted to simple sugars (gCH_2Om^{-2}, Loomis and Williams 1963), the quantity of nitrogen or water taken up by roots, the mass of leaves eaten by an herbivore or prey by a carnivore, or the number of hosts parasitized by a prasitoid. As we shall see, the functions describing them have similar shapes.

Type I Functional Response

Type I models such as the Lotka–Volterra (Lotka 1925; Volterra 1931), Nicholson and Bailey (1935), and Hassell and May (1974) models assume for a given predator density P that mortality increases linearly with prey density. This is easily seen in the L–V functional response for a simple model of a prey (N) − predator (P) interaction [3].

$$\frac{dN(t)}{dt} = rN - sPN \qquad [3a]$$

$$\frac{dP(t)}{dt} = hsPN - \mu P \qquad [3b]$$

In this model, r is the per capita prey birth rate, the quantity $0 < g(\cdot) = sN$ is the number of attacks per predator, and s is a constant of proportionality. As in [1], h is the rate of conversion of prey captured to new predators and μ is a constant predator death rate. Of course, the parameter s is a simplification of complicated interaction behavior (see below), but it might give a very reasonable description of predator success at low prey and predator densities or where little overlap of predator search efforts occur. Model [3] is unrealistic at high-prey densities because it implies that predators are insatiable and it does not account for search overlap by predators (Fig. 5.2a vs. 2b).

The constant s in many functional response models is viewed as the proportion of the universe searched per predator per unit time. This parameter is difficult to measure for real systems, but it can be estimated from field data under certain conditions (e.g., O'Neil 1988, 1989, 1990). For example, if N prey are offered to a predator in a given universe, and N_S prey survive, then the effective proportion of the universe covered during the study period T is $s = 1 - N_s/N$. Of course, the population of predators search some areas more than once, and as predator numbers increase, the degree of overlapping search will greatly increase unless they perceive previously searched areas (cf. Varley et al. 1973). The type I Nicholson–Bailey (N–B) model [4] predicts this kind of overlapping search behavior

$$N_a = N(1 - e^{-sPT})$$ [4]

where $(1 - e^{-sPT})$ is the "area traversed" by all predators, the exponent sP is a dimensionless number because s is defined as a per capita rate per unit time T, and P and N are per unit area. The term e^{-sPT} is roughly the zero terms of a Poisson distribution (i.e., the proportion of the area not searched). The shaded area in Figure 5.2b is the area searched more than once by the same or other predators. Note that [4] is linear in N, and hence is type I. In certain cases, the per capita rate of search s may be estimated by fitting [4] or a more appropriate model to the data. In fact, s is a per capita k-value (see Chapter 3; Varley et al. 1973).

$$s = \left(-\log_e \frac{N - N_a}{N}\right) \Big/ PT.$$ [4a]

The biology of search, predator and prey behavior, prey distributions, predator and prey preferences, age effects on attack performance, hunger, synomones, pheromones and kairomones, habitat characteristics, and sundry other

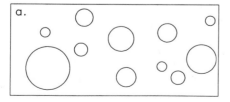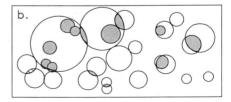

Figure 5.2. Types of search behavior: (a) nonrandom search and (b) random search. The shaded area in b is due to search overlap.

biotic and abiotic factors affect aquisition success (e.g., Vinson 1981; Weseloh 1981). This intuitive understanding of this complexity is part of the common wisdom, but unfortunately, incorporating these effects in a meaningful way in an analysis of the dynamics of species interactions has proven difficult. The role of mutual interference has probably been over emphasized (Free et al. 1977), being important only at very high-predator densities which occur only infrequently in the field.

Type II Functional Response

Holling's (1959, 1966) classic papers on the components of predation behavior set the stage for including the complexities of animal behavior in functional response models. Holling (1966) type II "disc equation" model has special historical interest as it was the first model to include handling time (T_h). This model is related to the Michaelis–Menten enzyme kinetics model (see Real 1979), and can be derived from the L–V model by assuming that the time for searching (T_s) equals the total time (T) minus the time spent handling prey captured (T_h) (i.e., $T_s = T - T_h N$). Substituting this definition in the L–V model yields

$$N_a = sPN(T - T_h N_a) \qquad \text{[5a]}$$

and rearranging terms gives us the Holling model for $P = 1$

$$\frac{N_a}{N} = \frac{sT}{1 + sT_h N}. \qquad \text{[5b]}$$

Model [5b] does not account for the fact that the available prey decrease during the time interval T, and hence it is applicable during very short intervals of time or possibly to situations where the natural enemies search systematically and do not reencounter previously exploited areas (Royama 1971; Rogers 1972; Hassell 1978). Royama recognized that predator and parasitoid functional response models were intrinsically different, and outlined the underlying theory (Appendix 5.1).

[As an aside, Holling's (1966) attempt to quantify and incorporate the effects of handling time and other aspects of individual behavior in the functional response model was premature because the hardware and software required for this amount of detail at the population level had not been developed. Some recent advances in object oriented programming (Markala et al. 1988, Huston et al. 1988, DeAngelis and Gross 1992), plus the increasing speed and power of microcomputers make Holling's vision a reality.]

Rogers (1972) (see Royama 1971) reformulated Holling's model to account for differences in parasite and predator attack biologies (with and without replacement, respectively), but still retained handling time. What this means is that parasitoids can superparasitize, but a prey can be eaten only once. Roger's models are

$$N_a = N\{1 - \exp[-s(PT - T_h N_a)]\} \qquad \text{predator [6a]}$$

$$N_a = N\{1 - \exp[-sPT/(1 + sT_h N)]\} \qquad \text{parasitoid [6b]}$$

Note that [6b] but not [6a] is ratio dependent. In all future discussions, T will be assumed equal to 1 and omitted from the equations.

A type II model based entirely on Poisson probability considerations (random search) is the Thompson's (1939) parasitoid model [7] with parameter E being the parasitoid per capita eggs supply (i.e., the demand for hosts).

$$Na = N(1 - e^{-EP/N}) \qquad [7]$$

In this model, the probability of escaping attack is exactly the zero term of a Poisson distribution (i.e., $\text{Prob}\{0\} = e^{-EP/N} = \exp(-$ total demand/total supply) and the probability of a host being attacked at least once is $[1 - f(P/N)]$ (i.e., the sum of the nonzero terms). The Thompson model serves to introduce ratio dependent models.

Ratio Dependent Functional Response Models [g(P/N)]. The simplest ratio dependent model is P/N (Berryman 1992), but other ratio dependent models are modifications of the Holling model (Getz 1984; Ginsberg and Akçakaya 1992), and the demand driven hunting equation model ([8a], Fig. 5.3) of Gutierrez et al. (1981a) and Klay (1987) (see Berryman et al. 1995 for a

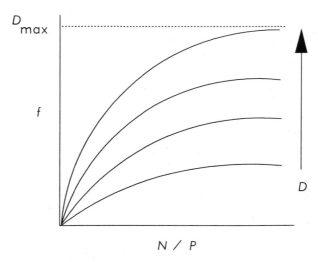

Figure 5.3. The Frazer–Gilbert and Gutierrez–Baumgärtner functional response models as a function of per capita consumer demand (D) and the ratio of N/P.

review). Because of the importance of the Gutierrez–Baumgärtner model (i.e., G–B models) to the theme developed in this book, it is discussed in some detail.

$$N_a = g(N/P) = DP[1 - e^{-\alpha N/DP}] \qquad [8a]$$

The parameter α in [8a] is the apparency rate or the proportion of the prey available for attack, and differs from s in other models—it is not Nicholson's *area of discovery* (Nicholson and Bailey 1935). It normally takes values from zero to unity, but the value may vary with species, organism age and size, responses to prey kairomones, preference, refuges, and with changes in the search universe. In practice, it is difficult to estimate α precisely for animals, but for plants it is known to increase with leaf area and is easily measured (Beer's law, Monteith 1965). For animals, it may be viewed as the probability that a hungry predator (or parasitoid female with a high egg load) will find a prey item at low density. However, if the predator uses kairomones to find prey, then the apparency and the rate of encounters might be considerably higher (e.g., Gutierrez et al. 1993). The quantity $1 - \alpha$ may be viewed as a prey refuge.

The per capita demand D increases realism by providing the motivating factor for search, limits consumption to realistic levels and extends the applicability of the model to a wide range of possible prey–consumer ratios (Gutierrez et al. 1984a, 1987, 1988a–c, 1994; see Chapter 6). The demand may be in number, energy, or biomass units.

Parasitoid and predator forms of this model can be derived by making simple assumptions about random search and attack behavior (see Appendix 5.2). The parasitoid form (reinterpreted in a fully ratio dependent form as [8b]) was first proposed by Frazer and Gilbert (1976), where D is the per capita demand rate for hosts per unit time, α is the proportion of hosts that may be attacked, and N and P are defined above.

$$N_a = N\left[1 - \exp\left\{\frac{-DP}{N}\left[1 - \exp\left(\frac{-\alpha N}{DP}\right)\right]\right\}\right] \qquad [8b]$$

The model approaches a constant as N goes to zero. This can be demonstrated by taking a Taylor's expansion of the second exponential $(1 - e^{-\alpha N/DP})$ in [8b] as $N \to 0$.

$$1 - 1 + \frac{\alpha N}{DP} - \frac{(\alpha N/DP)^2}{2!} + \frac{(\alpha N/DP)^3}{3!}$$

$$- \cdots + \frac{(\alpha N/DP)^{(n-1)}}{n-1!} + \frac{(\alpha N/DP)^n}{n!}$$

$$[8c]$$

Ignoring the higher order terms yields $\alpha N/DP$, which when substituted simplifies [8b] to $N(1 - e^{-\alpha})$. This suggests that at very low prey densities, the parasitoids attack a constant proportion of the prey not in a refuge. However, at high levels of hosts, the function saturates to the maximum population demand (DP). We shall return to the G–B model several times in this book.

Type III Models

It is commonly accepted that the efficiency of resource acquisition by vertebrate predators increases with resource density and that their functional response is characteristically type III. This has commonly been thought to indicate learning, and has been demonstrated in invertebrate predators. Specifically, Hassell (1978) showed that a type II model becomes type III when the predator search rate is a concave function of prey density [e.g., $s(N) = cN/(1 + dN)$ with constants c and d]. Hassell's modifications of Roger's models is an example (see Table 1).

Real (1979) reanalyzed Holling's experiments on deer mouse *Peromyscus maniculatus* predation on a preferred prey (European sawfly pupae) and a less palatable one (wheat seeds) to demonstrate learning in vertebrates. He showed the effects of food choice and distribution on the transition of a type II functional response to a type III sigmoid response. Real's model is

$$N_a = \frac{KA^\phi}{X + A^\phi} \qquad [9]$$

where N_a is the feeding rate, K is the feeding rate at saturation, A is the density of food items, ϕ is the parameter associated with the increase in the rate of detection of food items with increasing density (i.e., learning by the organism), and X is the food density that generates one-half the maximum feeding rate. If $\phi = 1$, the function is type II and reduces to Holling's model. If $\phi > 1$, [9] is type III and measures the degree to which the rate of detection is influenced by learning in response to increasing prey density, but $\phi < 1$ suggests a degree of avoidance. Rearranging [9] and taking the logarithm of both sides linearizes it:

$$\log \frac{N_a}{K - N_a} = \log \frac{1}{X} + \phi \log A \qquad [10]$$

Estimates of ϕ and X can be obtained by regressing $\log N_a/(K - N_a)$ on $\log A$.

The analysis showed that deer mice ate more sawfly cocoons when given in combination with low-palatable alternative food than when presented alone, but the learning coefficient ϕ was the same Figure 5.4a and b. In contrast, consumption rates were lower and learning greater when prey items were randomly dispersed rather than clumped (Figs. 5.4c and d).

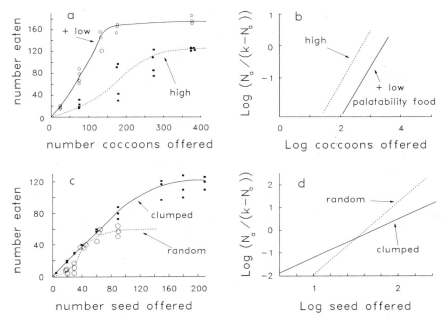

Figure 5.4. *Real's analysis of prey composition and distribution on predator learning: (a) Holling's data on deer mice, (b) plot of log $N_a/(K - N_a)$ on log sawfly cocoons plus high- and low-palatability alternate prey (see text), (c) the effects of spatial distribution of wheat seed on learning, and (d) (see B above). [Reprinted with permission from Real 1979.]*

ESTIMATING THE PARAMETERS OF FUNCTIONAL RESPONSE MODELS

Satisfactory fits of functional response models to data from laboratory feeding experiments are common (e.g., DaSilva et al. 1992), and some success has been obtained using field experiments (see O'Neil 1988, 1989), but in general the results are mixed. For example, in Gilbert's (1981) study of coccinelid predation on the pea aphid, the aphids fell off the plant as the coccinelid predators approached. This escape behavior reduced predation success and quite possibly increased prey survival, and proved difficult to quantify and incorporate in the model. Greater success has been obtained with sessile species such as in Williams' (1984) study of parasitism of blackberry leafhopper [*Dikrella californica* (Lawson)] eggs (N) by the short-lived parasitoid *Anagrus epos* Girault (P). The field data proved relatively easy to fit using a Holling–Michaelis–Menten type II model, [11]. Leafhopper eggs are imbedded in leaf tissue on the under side of the leaf, hence rainfall and other factors that cause serious disruptions in other systems (Hassell 1986; Gutierrez et al. 1988b) are largely absent. In addition, parasitized eggs [$P(t)$] are easily distinguished from unparasitized eggs [$N(t)$]. The mean developmental time (τ) of the parasitoid

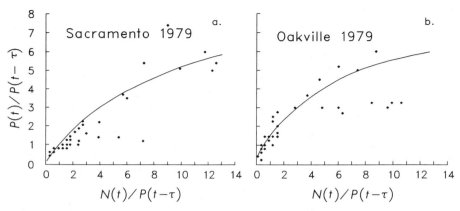

Figure 5.5. *The type II functional response of* A. epos *to blackberry leaf hopper eggs. [Reprinted with permission from Williams 1984.]*

egg to the adult stage was determined and used to extrapolate from the field data the number of parasitoid females that emerged from eggs $[P(t - \tau)]$ parasitized at time $t - \tau$. These data were used to plot the observed per capita attack rate [i.e., $N_a = P(t)/(P(t - \tau)]$ on the host/parasitoid ratio $[N(t)/P(t - \tau)]$ for populations at two locations (Fig. 5.5).

$$N_a = g(N, P) = \frac{0.47N(t)/P(t - \tau)}{1 + 0.47N(t)/P(t - \tau)} \qquad [11]$$

Note that $P(t - \tau)$ is in both the numerator and the denominator in [11]. Note also that any type II functional response model could be fit to the data, but only the biology determines which should be used.

ESTIMATING PREY PREFERENCE

Several models have been developed to quantify host preference or selective predation in natural systems and in the laboratory (Ivlev 1955; Murdoch 1969; Manly et al. 1972; Chesson 1983). The common methods of studying preference are outlined by Cock (1978). Figure 5.6 graphically illustrates these notions for preferred prey N and less preferred prey N', respectively, where N_a and N'_a, respectively, denote the quantity eaten or attacked. Incorporating preference into functional response models is covered in Hassell (1978) and others, and not discussed here. The form depends greatly on the functional response model used (see Schreiber and Gutierrez 1996 and Berryman et al. 1995).

The most widely used index of preference is Murdoch's index, which has been proposed at least six times (Cain and Sheppard 1950; Kettlewell 1956; Tinbergen 1960; Murdoch 1969; Elton and Greenwood 1970). All of these

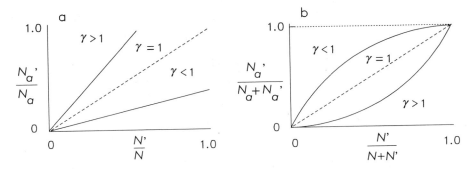

Figure 5.6. *Two methods for illustrating preference.*

methods can be mathematicaly related for two prey systems, and many can be mathematically related for systems having three or more prey species. Murdoch's index for a two host system is written as

$$\frac{N'_a}{N_a} = \gamma \frac{N'}{N} \qquad [12]$$

where $0 < \gamma < 1$. This method is useful only when the level of prey exploitation without replacement is negligible. To allow for exploitation, the Manly et al. (1972) index is more suitable as it is derived using instantaneous relationships. If p' is the probability that the next prey eaten is of type N', and γ is the preference coefficient, then the instantaneous relationship is

$$p' = \frac{n'_a}{n_a + n'_a} = \frac{n'}{n + \gamma n'} \qquad [13]$$

where n_a, n'_a, n, and n' are the instantaneous equivalents of N_a, N'_a, N, and N'. Rearranging [13] yields Murdoch's index for the case when prey exploitation is negligible. Manly et al. (1972) derived a probabilistic solution for cases when exploitation is important but prey density is large ([14], see the original paper for details).

$$\gamma = \frac{\log_e (N/N_a)}{\log_e (N'/N'_a)} \qquad [14]$$

Cock (1978) showed how preference can be predicted from functional response models fitted to data. For example, the Holling model can be adapted for two prey types to give

$$N_a = \frac{sNT}{1 + sT_hN + s'T'_hN'} \qquad [15a]$$

$$N'_a = \frac{s'N'T}{1 + sT_hN + s'T'_hN'}, \tag{15b}$$

s and s', and T_h and T'_h are the predator search rates and handling times for prey type N and N', respectively. Note that dividing [15a] by [15b] yields

$$\frac{N_a}{N'_a} = \frac{sN}{s'N'} \tag{16}$$

or Murdoch's index with $\gamma = s/s'$ (see Fig. 5.6a). Similarly, the instantaneous version of Holling's model for preference for two prey types (Lawton et al. 1974) yields this relationship.

$$\frac{N_a}{N'_a} = \frac{\{1 - \exp[-s(T - T_hN_a - T'_hN'_a)]\}N}{\{1 - \exp[-s'(T - T_hN_a - T'_hN'_a)]\}N'} \tag{17}$$

This result is also analogous to Murdoch's index when handling times T_h and T'_h approach zero. In this case, there is little difference between models that ignore or include the influence of exploitation and preference, which if it exists, is part of the discovery rate parameters s and s'. If handling time is significant, the interpretation of [17] is difficult.

A general preference model $[0 < \gamma (N, N') < 1]$ that reflects preference for the less preferred host N' in relation to the preferred host N should behave under extreme conditions as follows:

$$\gamma(N, N') \to 1 \quad \text{as} \quad N \to 0 \tag{18a}$$

and

$$\gamma(N, N') \to \varepsilon > 0 \quad \text{as} \quad N' \to 0 \tag{18b}$$

where $0 < \varepsilon < 1$ may reflect some nutritional requirement for N'. For example, big-eyed bugs (*Geocoris* spp.) feed on plant sap to meet maintenance requirements, but require arthropod prey for reproduction.

Theoretical relationships between prey eaten of type N (i.e., N_a) and prey eaten of type N' (i.e., N'_a) that satisfy conditions [18], do not necessarily satisfy the condition $\{N'_a(t) < N'(t) \text{ and } N_a(t) < N(t)\}$, especially at low densities of the preferred prey N and high densities of nonpreferred prey N'. Note that [12] may fail these conditions when $N'_a \to N'$. This result is easily seen by rearranging [12] as follows:

$$\frac{N'_aN}{N'N_a} = \gamma \tag{18c}$$

and noting that $\gamma < 1$ requires that $N_a > N$ which of course is impossible.

Predators are more likely to exhibit prey preferences when food is not limiting. This notion is depicted in Figure 5.7, where the preference for N' over N is plotted on predator food supply/demand ratio for weak-to-strong preference. All preference models may fail under extreme conditions of hunger or lack of hosts when host preferences may break down.

ESTIMATING PREFERENCE IN THE FIELD—SOME EXAMPLES

Pickett et al. (1987, 1989) utilized a variant of Chesson's (1981, 1983) and the Wilson and Gutierrez (1980) models to estimate the relative preference shown by biotypes of the leafhopper egg parasitoid $A.$ *epos* for grape and variegated leafhoppers. Both equations are similar, but in the former the sum of its preference coefficients equals 1.0, and in the latter preference is scaled to the maximum preference. Applied to the Pickett et al. (1989) example, the models have the following forms:

$$r_i = \frac{C_i N_i}{\sum\limits_{i=1}^{2} C_i N_i} \quad [19]$$

where r_i is the proportion of eggs parasitized by a particular biotype that belongs to the ith leafhopper, and C_i is the relative preference shown by a biotype for the ith leafhopper. Chesson's (1983) measure of preference for variegated leafhopper (γ_1) is calculated using [20]

$$\gamma_1 = \frac{r_1}{n_1} \bigg/ \left[\left(\frac{r_1}{n_1} \right) + \left(\frac{r_2}{n_2} \right) \right] \quad [20]$$

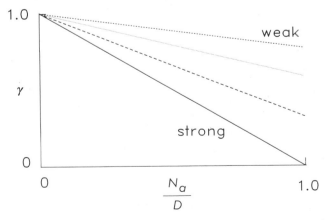

Figure 5.7. The hypothetical effects of food supply–demand (= N_a/D) on preference (γ) for alternate prey (N').

where r_1 = the proportion of parasitized eggs that are variegated leafhopper, n_1 = the proportion of all eggs that are variegated leafhopper eggs, r_2 = proportion of parasitized eggs that are grape leafhopper, and n_2 = the proportion of all eggs that are grape leafhopper eggs. Preference for grape leafhopper is $\gamma_2 = 1 - \gamma_1$, and the values vary from 0 (no preference) to 1.0 (maximum preference) and are scaled (normalized) such that the r_i values sum to 1.0. These preference values are therefore a relative measure and indicate the preference of a parasite for a host in relation to other hosts present.

The Wilson and Gutierrez (1980) measure of preference (γ_i) is calculated for leafhoppers using [21]

$$\gamma_i = \frac{r_i}{n_i} \bigg/ \max\left(\frac{r_i}{n_i}\right) \qquad [21]$$

where r_i = the proportion of parasitized leafhopper eggs that are of type i, n_i = the proportion of all eggs that are of type i, and max (r_i/n_i) is the maximum ratio of r_i/n_i for all prey types.

Wilson and Gutierrez (1980) developed this empirical preference model from field data, and used it to estimate not only preference but also handling times of different aged cotton bollworm (*Heliothis zea* Hubner) larvae feeding on different aged cotton fruits. In the study, larvae of known ages were observed searching cotton plants in the field for populations of fruits of varying ages throughout the season. Bollworm larvae, like many predators (herbivores) exhibit changing preferences, especially as the age distributions of both the predator and prey and their strata within the plant change. The applicability of the model would be severely limited if parameters had to be calculated for each time and circumstance. This proved not to be the case as in 20 of 23 large data sets the data conformed to the predictions of the model.

EPILOGUE

The success of an organism in resource acquisition is influenced by many biotic and abiotic factors, and the aggregate of individual performances determines the growth dynamics of the population. The treatment of this topic here has been an average or per capita one. We do not know whether this average effect is the same as one would get if individuals were modeled as they interact with prey and with each other. This would require us to study individual behavior and develop a model of individual success (e.g., object oriented models (see Chapter 10)), and then to aggregate the results at the population level (e.g., Casas 1989).

Predator and Parasitoid Forms of Functional Response Models (see Royama 1971)

The instantaneous forms of predator and parasitoid functional response models can be derived from basic assumptions about hunting. What distinguishes the two forms is that predators (Y) remove the prey (X) making them unavailable to other predators and that several parasitoids (also Y) may attack the same host more than once. Royama (1971) began with the assumption that the instantaneous per capita predator search (s) is constant [i.e., $f(x) = sx$], and parasitoids do not discriminate among previously attacked hosts (i.e., the attacks are random—Prob$\{0\} = e^{-n/x}$). Using these notions and the basic hunting equation $n = sxYt$ as starting points, he derived the instantaneous predator and parasitoid forms of the functional response model. Note n is the number of attacks, x is the varying number of prey during Δt, and t is time.

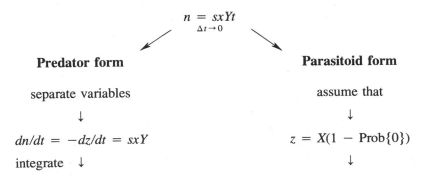

$$n = sxYt$$
$$\Delta t \to 0$$

Predator form

separate variables

↓

$$dn/dt = -dz/dt = sxY$$

integrate ↓

Parasitoid form

assume that

↓

$$z = X(1 - \text{Prob}\{0\})$$

↓

$$\int_{x_o}^{x} - dx/x = \int_{0}^{t} sY dt \qquad\qquad \mathrm{Prob}\{0\} = e^{-n/x}$$

$$\downarrow \qquad\qquad\qquad\qquad \downarrow$$

$$\qquad\qquad\qquad\qquad\qquad \text{substituting for } n$$

$$-\ln(x/x_o) = sYt \qquad\qquad \mathrm{Prob}\{0\} = e^{-sxYt/x}$$

$$\downarrow \qquad\qquad\qquad\qquad \downarrow$$

$$x/x_o = e^{-sYt} \qquad\qquad \text{and since } X = X_o$$

$$\downarrow \qquad\qquad\qquad\qquad \downarrow$$

$$x = x_o e^{-sYt} \qquad\qquad z = X_o(1 - e^{-sYt})$$

$$\downarrow$$

$$n = x_o - x \rightarrow \qquad \text{number attacked} \qquad \hookleftarrow$$

Note that by substitution the predator and parasitoid forms of the model are equivalent.

If, say, the Gutierrez–Baumgärtner per capita hunting equation model is used as our starting point $[f(x) = D(1 - e^{-\alpha x/DY})]$, the resulting predator and parasitoid forms are quite different. In the model, search is random, prey availability or apparency is α, and the model predicts how much of the demand (D) for prey will be met at given prey and predator densities. If the area to be searched is small or the time large, α may be quite large.

The Predator Form of the G–B Model

$$\frac{dn}{dt} = -\frac{dx}{dt} = D(1 - e^{-(\alpha x/DY)})Y$$

$$-\frac{dx}{DYdt} = (1 - e^{-(\alpha x/DY)})$$

$$-DYdt = \frac{dx}{(1 - e^{-\alpha x/DY})}$$

Integrating the left-hand side yields

$$-DYt = \int_{x_o}^{x} \frac{dx}{1 - e^{-(\alpha x/DY)}}$$

and then the right-hand side

$$= \frac{DY}{\alpha} \ln \left[e^{\alpha x/DY} - 1\right]\big|_{x_o}^{x}$$

$$= \frac{DY}{\alpha} \ln \left[\frac{(e^{\alpha x/DY} - 1)}{(e^{\alpha x_o/DY} - 1)}\right]$$

and simplifying and taking the antilog of both sides

$$e^{-\alpha t} = \frac{e^{-(\alpha/DY)(x_o - x)} - e^{-(\alpha x_o/DY)}}{1 - e^{-(\alpha/DY)x_o}}$$

$$-\frac{\alpha}{DY}(x_o - x) = \ln\left[(1 - e^{-(\alpha/DY)x_o})e^{-\alpha t} + e^{-(\alpha/DY)x_o}\right]$$

Finally, since x_o equals the initial x, then

$$n = (x_o - x) = -\frac{DY}{\alpha}\ln\left\{[1 - e^{-(\alpha/DY)x}]e^{-\alpha t} + e^{-(\alpha/DY)x}\right\}.$$

The Parasitoid Form. We can similarly derive the Frazer–Gilbert parasitoid model from the basic G–B hunting model. If $n = sxYt$, then substituting the G–B model for n in the general hunting equation:

$$z = X(1 - e^{-n/X})$$

and rearranging terms yields the more familiar form of the F–G model (Gutierrez and Wang 1977).

$$z = X\left\{1 - \exp\left[-D\frac{Y}{X}\left(1 - \exp\frac{-\alpha X}{DY}\right)\right]\right\}$$

Theoretical Basis for the Instantaneous Gutierrez–Baumgärtner Hunting Model (G–B)

Definitions

N_a is the number attacked during an infinitesimally small period of time dt
N_o is the initial number of prey that is minimally depleted during dt
D is the physiological demand rate for prey per predator during dt
P is the number of predators
A is the effective search rate per P

The model (G–B) proposed by Gutierrez et al. (1981a) is similar to those of Ivlev (1955) and Watt (1959) (see Royama 1971 for a review), but in the G–B model random search is driven by the physiological demand for resource, a refuge for prey is a component of the model, and all of the parameters have measurable physical or behavioral meaning.

Assume a general model of the form

$$\frac{\partial N_a}{\partial N_o} = f(N_o, P) \tag{A5.1}$$

and as in Royama (1971)

$$\frac{\partial N_a}{\partial N_o} = f(N_o, P) = AP(DP - N_a) \tag{A5.2}$$

This model differs from Watt's model in that $\partial N_a/\partial N_o$ declines as N_a approaches DP (i.e., the predator is satiated). The effective rate of search A, however, should decrease as $P \to \infty$ because of increased competition.

$$\frac{dA}{dP} = -\frac{\beta A}{P} \tag{A5.3}$$

where β is the coefficient of competition. Integrating [A5.3], yields

$$A = sP^{-\beta} \tag{A5.3.a}$$

with search coefficient s, which means that as $P \to \infty$ then $A \to 0$. Substituting [A5.3.b] in [A5.2] we get

$$\frac{\partial N_a}{\partial N_o} = sP^{1-\beta}(DP - N_a) \tag{A5.4}$$

Integration [A5.4] is accomplished by first separating the variables

$$\frac{\partial N_a}{(DP - N_a)} = sP^{1-\beta}\partial N_o$$

and integrating both sides of the equation as follows:

$$\int \frac{\partial N_a}{(DP - N_a)} = \int sP^{1-\beta}\partial N_o$$

$$-\ln(DP - N_a) = sN_o P^{1-\beta} + C$$

$$DP - N_a = e^{-sN_o P^{1-\beta} + C}$$

If we assume that $e^C = DP$ and $s = \alpha/D$, then

$$N_a = DP(1 - e^{-(\alpha/D)N_o P^{1-\beta}}) \tag{A5.5.a}$$

This result is very similar to Watt's model, but asks a very different question; namely what proportion of the demand for prey was met given N_o and DP, where

$$0 \leq \frac{N_a}{DP} = (1 - e^{-(\alpha/D)N_o P^{1-\beta}}) \leq 1 \tag{A5.5.b}$$

is the proportion of the demand met (i.e., the supply/demand ratio) and $e^{-(\alpha/D)N_o P^{1-\beta}}$ is the shortfall. However, as Royama (1971) points out, this model has some difficulties concerning the competition parameter β as $P \to \infty$.

$$N_a = \lim_{P \to \infty} f(N_o, P)$$

$$= \begin{cases} \text{(case 1) if } 0 \le \beta < 2, N_a \to \infty & \text{(facilitation)} \\ \text{(case 2) if } \beta = 2, N_a \to DP & \text{(intraspecific competition)} \\ \text{(case 3) if } \beta > 2, N_a \to 0 & \text{(hyper intraspecific competition)} \end{cases}$$

[A5.6]

Case 1: If N_o is small, N_a may be larger than N_o because β acts as a facilitation parameter. To correct this DP must be less than αN_o because $1 - \alpha$ is the proportion of prey in a refuge.

Case 2: This is the normal model and has no obvious mathematical pathologies.

Case 3: In the extreme, hypercompetition as $P \to \infty$ may become so severe that $N_a \to 0$.

In summary, the model [A5.5] incorporates random demand-driven search and all parameters have measurable physical and/or behavioral meaning, and intuitive reasonable limits must be applied to the parameter β.

6

Resource Acquisition and Allocation

It would be instructive to know by what physiological mechanism a just apportionment is made between the nutriment devoted to the gonad and that devoted to the rest of the parental organism, but also what circumstances in the life history and environment render profitable the diversion of a greater or lesser share of the available resources toward reproduction.
—R. A. Fisher, 1930

The currency of trophic interactions is of utmost importance in the development of population models. Historically, the numbers of interacting organisms have been used in models, but Lotka (1925) knew energy, biomass, or other units could be equally valuable. In this chapter, we examine resource acquisition (see Chapter 5) and allocation to growth, development, and reproduction. The model is easily extended to the myriad of possible biologies. My goal is to lay the foundation for the development of physiologically based age—structured multitrophic population dynamics models.

The pioneering works of Phillipson (1966), Odum (1971), Weigert (1976), and others on ecological energetics, and the readable summary of the field found in Townsend and Calow (1981), and Llewellyn (1988) provide important background on this area. The relationships of size to growth, respiration, and other physiological rates made by Peters (1983) prove invaluable to ecosystems studies. I am cognizant of the special terminology that is part of ecological energetics research (Scriber and Feeny 1979; Llewellyn, 1988), but I take liberty in simplifying the notation and in interpreting observed allocation patterns to include these concepts in plant and animal (arthropod) population models.

SOME IMPORTANT ANALOGIES

All organisms must address the problems of resource acquisition (photosynthesis in plants or feeding in animals) and allocation to egestion (e.g., carnivorous plants), respiration, and costs of coverting biomass from lower trophic levels to growth and reproduction. These two processes are the *functional* (acquisition) and *numerical* (allocation) *responses*, respectively. Although species may differ with regard to biological details, the shapes of the resource acquisition functions at all trophic levels are similar (type II or III, see Chapter 4), and the allocation of the resources have the same priorities (Fig. 6.1, cf. Gutierrez and Curry 1989). All of these physiological and behavioral activities have costs associated with them. In poikilotherms, the cost of obtaining a

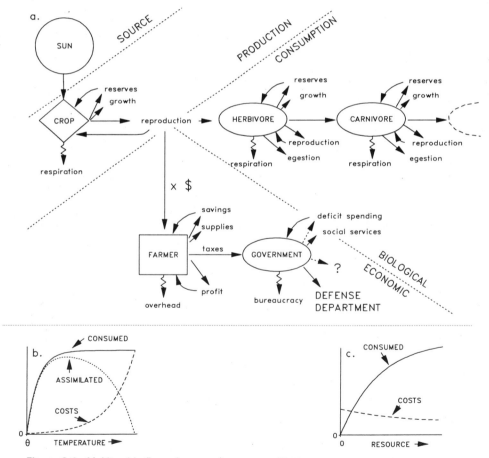

Figure 6.1. *Multitrophic flow of energy (a, see text (Gutierrez and Curry, 1989)); (b) the acquisition, cost, and assimilation functions on temperature; and (c) the acquisition and cost functions on resource availability.*

resource increases with temperature, and in most organisms it decreases with increasing resource availability (Fig. 6.1b and c). The difference between the resource acquired (i.e., benefits) and the cost of acquiring it is the amount of energy that can be fixed as growth, reproduction, or reserves. In economic models, the benefit or resource acquisiton function has the same yield–effort relationship, and the costs of resource acquisition increase as the effort required to obtain them increase. Only at the governmental level do these analogies break down as there is no obvious analog for biological fitness (i.e., reproduction or profit) unless the production of bureaucrats qualifies as reproductive output.

Plants (*autotrophs*) and animals (*heterotrophs*) are all predators by our broad definition (Chapter 1), though to be more precise, they may be predacious (consume the resource making it unavailable to others) or parasitic (the same host may be attacked by others), or they may be one or the other during part of their life or a mixture of both. Resource acquisiton by autotrophs and heterotrophs differs in two important respects: autotrophs seek inorganic nutrients such as nitrogen, as well as water and light, which through the process of photosynthesis are used to form the complex molecules that construct self. Heterotrophs (including carnivorous plants) seek prey composed of complex molecules that are digested, and the components assimilated and reassembled to form self. Some animals are polyphagous and may search for different kinds of resources to balance their diet, but adequate nesting sites, mates or territories may also be viewed as necessary resources for animals that affect their vital rates.

That individual plants have populations of subunits within them was popularized by Harper and White (1974; see Gutierrez et al. 1975), and similarly, individual animals in a population may also have a population of reproductive subunits within them. Gutierrez and Wang (1977) suggested that a common population dynamics model could be used for the subpopulations within individuals and for populations of parent individuals themselves, and that the population model's dynamics could be regulated by the interplay of resource acquisition and demand. This theme is developed here and in subsequent chapters.

DEVELOPMENTAL PHENOLOGY

The importance of growth and reproductive strategies (Gilbert 1990) in the dynamics of resource allocation is also a major theme of this chapter. The variations are nearly limitless, but despite the apparent biological differences, analogies exist and can be exploited in the development of individual and population level models.

Vegetative growth in plants is analogous to immature growth in animals, and seed production is analogous to progeny production. The growth phenology of some plants is highly deterministic, some have distinct cyclic periods of vegetative growth followed by a period of reproduction, while in others both may occur simultaneously (i.e., indetermine growth). Age specific *demands* of reproductive subunits may be small or large relative to the potential resource acquisition rate (*supply*) of the parental organism, and as we shall see this interaction greatly affects determinism.

In some animals, ova mature within the parent where they are fertilized and are oviposited. In viviparous aphids, for example, ova may not require fertilization and may continue rapid growth within the female until deposited as walking nymphs (Fig. 6.2). Parasitoids may be predacious as immatures and have their full complement of eggs at maturity, and the adult demand is the

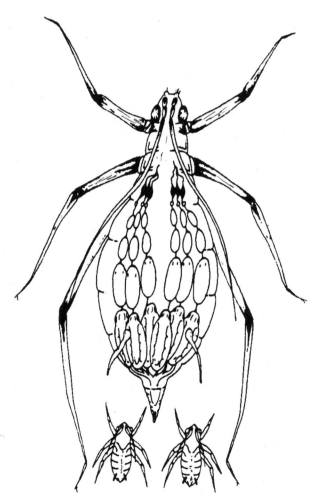

Figure 6.2. *An aphid illustrating the concept of populations within populations. [Reprinted with permission from Dixon 1987.]*

number of hosts that may be attacked. In other cases, the demand may also include hosts for host feeding. In some insects, padeologic reproduction may occur [i.e., reproduction by immatures, Chapman 1982, and in some growth may be cyclic or indeterminate (spiders, crabs, etc., Sebens 1987)].

The production of storage organs in plants is analogous to the accumulation of fat reserves in animals. Plants and animals may use reserves during periods of resource shortfall, dormancy, or stress.

The interplay between the rates of resource acquisition (supply) and organism genetic demands (i.e., *the supply/demand ratio*) in the model determines the phenology of allocation priorities to and the realized rates of growth, reproduction, and survivorship of whole organisms and of the subunits they might contain. Abiotic and biotic factors may affect the demand or the supply side of this ratio, and either may greatly affect the dynamics of growth and development, and in the aggregate, they may affect the dynamics of the population (Gutierrez and Baumgärtner 1984b).

The processes of energy acquisition and allocation are interconnected and determine all aspects of organism growth and development. These processes are components of the concept (see Petrusewicz and MacFadyen 1970; de Wit and Goudriaan 1978) called the *metabolic pool* (Gutierrez et al. 1975, 1981a; Wang et al. 1977).

THE PER CAPITA METABOLIC POOL MODEL

The assimilation rate $[A(a, t, \tau)]$ of an organism of mass $M(a, t)$ of age a at time t held at temperature τ is the difference between the resource acquired $[S(t, \tau, \cdot) = the\ supply]$ and the basal respiration rate $[z(\cdot, \tau)]$ corrected for egestion (β) and tissue conversion costs (λ) (i.e., *all of the costs*). The notation ($\cdot$) simply means that the model has additional variables that will be defined below. This assimilate may be allocated to growth (G) and/or reproduction (R) (subscripts ignored, see Fig. 6.3). Obviously, the rate A is greatly influenced by the success organisms have capturing resources, the abundance of which may vary greatly over time. The dynamics of the metabolic pool model may be represented mathematically in discrete form [1].

$$A(t) = G(t) + R(t) = \{\beta S(\cdot) - z(\cdot)M(t)\}\lambda \qquad [1a]$$

The simplest difference model for assimilation (i.e., growth) of an individual of mass M at time t is

$$M(a + 1, t + 1) = M(a, t) + A(t) \qquad [1b]$$

(Gutierrez et al. 1981a; Gutierrez and Baumgärtner 1984a and b; and Baumgärtner et al. 1987). The assimilation rate A may vary as follows:

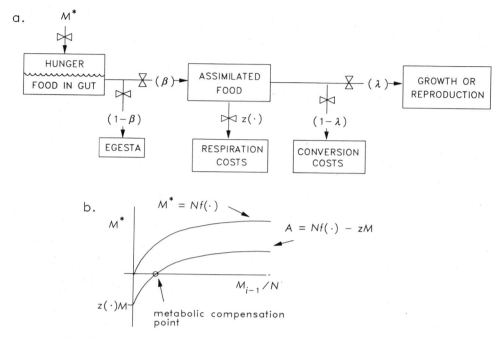

Figure 6.3. *The metabolic pool model (a) [Redrawn from Baumgärtner et al. 1987.] (b) The acquisition function indicating the metabolic compensation point. [Redrawn from Gutierrez et al. 1981.]*

$$A \text{ is } \begin{cases} negative & \text{if } \beta S < z(\cdot)M \text{ (i.e., negative growth)} \\ zero & \text{if } \beta S = z(\cdot)M \text{ (i.e., the metabolic compensation point)} \\ positive & \text{if } \beta S > z(\cdot)M \text{ (i.e., positive growth)} \end{cases}$$

Figures 6.4a and 6.5a elaborate on the components of [1a] and Figures 6.4b and c and 6.5b and c illustrate the the factors affecting resource acquisition and allocation in plants and animals, respectively.

The Acquisition Model [$S(\cdot)$]

The resource acquisition or functional response model $S(\cdot)$ includes consumer demand physiology, intraspecific competition, and demand driven random search ([2], Gutierrez 1992; see also Chapter 5).

$$S(N, P, t, \tau) = f\left(\frac{N}{P}\right) = (1 - e^{-\alpha N/DP\Delta t})D\Delta t \qquad [2]$$

By way of review, $S(\cdot)$ is the average per capita consumption rate of prey (N) per predator (P) in the face of intraspecific competition (i.e., the exponential part) at temperature τ over a time period Δt. The constant α is the *prey apparency rate* or the proportion of N available for attack during Δt (i.e., prey not in a refuge), and D is defined as the *genetic maximum* per capita demand rate for prey per unit of physiological time (see Chapter 3). The prey/predator ratio in the exponent makes this a random search model that includes the effects of intraspecific competition. In future discussions Δt is assumed to be unity and is ignored, though in practice it is temperature dependent and may vary considerably on a daily basis in nature. The parameters D and α must be estimated for each species.

The acquisition model may also be written in a mass dependent form for trophic level 2 preying on level 1.

$$S(M_1, M_2, t, \tau) = f\left(\frac{M_1}{M_2}\right) = (1 - e^{-\alpha_2 M_1/D_2 M_2})D_2 \qquad [3]$$

Here S is the average acquisition rate of resource mass (M_1) (i.e., supply) per unit of consumer mass (M_2), the masses of population P and N are $M_1 = Nm_1$ and $M_2 = Pm_2$, respectively, and m_1 and m_2 are the average mass, D_2 is per unit mass demand of M_2 per unit of physiological time, and α_2 is the proportion of the resource that is available to all consumers. This form is very convenient as the size of both predator and prey and the demand per unit mass vary over time (see below).

In theory, the supply of resource available for all physiological processes is always less than the total demand (e.g., $0 \leq S/D < 1$) even when resource is nonlimiting because of the inefficiencies of search (Gutierrez and Wang 1977). Shortfalls of other essential nutrients (or requisites) may also affect growth, reproduction, and, as we shall see, survival rates. Their effects are reviewed in a later section.

Estimating the Components of the Demand Rate (D)

The *genetic demand* rate D in $S(\cdot)$ is conceptually the sum of the genetic maximum outflow rates of the metabolic pool model corrected for metabolic and respiratory costs (i.e., β, z, λ, Fig. 6.3, see below). Under nonlimiting conditions it may be estimated by rearranging the metabolic pool model [1] to [4].

$$D > \frac{(A/\lambda) + z(\cdot)M}{\beta} \qquad [4]$$

It is easy to estimate the components from controlled laboratory experiments such as those published by Randolph et al. (1975) for pea aphid (*Acyrthosiphon pisum* Harris) (Fig. 6.6, Gutierrez and Baumgärtner 1984). Specifically, the

a. The metabolic pool model and allocation to plant parts

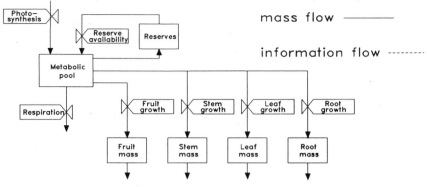

mass flow ─────────

information flow ─ ─ ─ ─ ─

b. Factors affecting photosynthesis

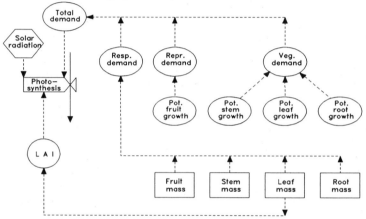

c. Supply−demand effects on allocation

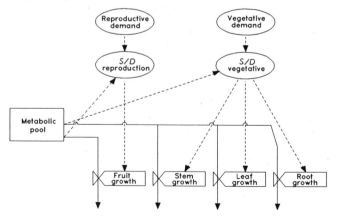

Figure 6.4. Components of the plant metabolic pool model. [Reprinted from Agric. Syst., 32, Graf et al., A simulation model for the dynamics of rice growth and development Part I—The carbon balance, pp. 341–365, copyright © (1990a), with kind permission from Elsevier Science Ltd., The Boulevard, Langford Lane, Kidlington OX5 1GB, UK.]

a. The metabolic pool model and allocation to animal subunits

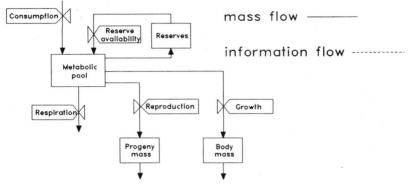

mass flow ————

information flow ------------

b. Factors affecting consumption

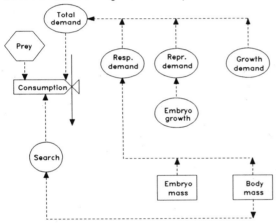

c. Supply–demand effects on allocation

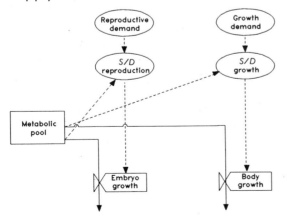

Figure 6.5. *Components of the animal metabolic pool model. [Modified from Graf et al. 1990a.]*

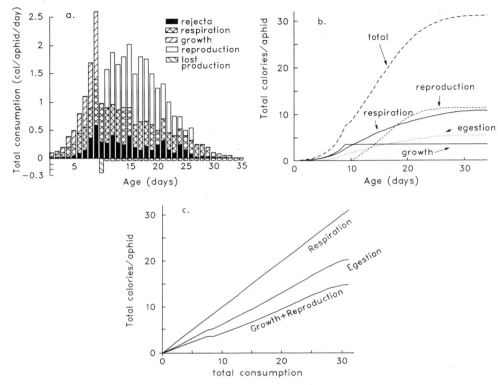

Figure 6.6. Energy allocation in pea aphid: (a) age-specific energy allocation [Reprinted with permission from Randolph et al. 1975], (b) cumulative allocation on age, and (c) cumulative allocation [Reprinted with permission from Gutierrez and Baumgärtner 1984a.]

proportion of the food consumed going to egestion, respiration, and growth plus reproduction are 0.176, 0.338, and 0.486, respectively (Fig. 6.6c). Note that the term for conversion efficiency (λ) is part of respiration. It is, however, considerably more difficult to estimate these parameters from field data because the environmental variables are usually not controlled. Because our main goal is to understand the dynamic of assimilation, we can accept reasonable estimates for the parameters.

Assimilation Demand for Growth and Reproduction (A). The mass of individuals of age $a + \Delta a$ can be plotted on their mass at age a and used to estimate the *compound growth rate* $(1 + \gamma)$ of an average individual of mass $M(a)$. Note that time and age may be in the same units.

$$M(a + \Delta a) = M(a)(1 + \gamma \Delta a) \qquad [5]$$

We estimate the instantaneous assimilation rate per unit mass per da by rearranging terms and dividing both sides by Δa

$$\frac{dM}{da} \approx \frac{M(a + \Delta a) - M(a)}{\Delta a} = \gamma M(a) \quad \text{as} \quad \Delta a \to 0 \quad [6]$$

The solution of [6] is the exponential growth model

$$M(a) = M_0 e^{\gamma a} \quad \text{for} \quad 0 \le a < a^* \quad [7]$$

where M_0 is the mass of a new individual born of age $a = 0$, and a^* is the age at reproductive maturity. Under nonlimiting conditions, $g(a) = \gamma e^{\gamma a}$ is the instantaneous age specific maximum per unit mass demand for growth. This simplified method for estimating age-specific $g(a)$ works quite well for animals, but less well for plants because plant parts abscise, greatly affecting the estimate. It is best to use young plants for this purpose.

The total growth demand (D_G) of an age-structured population of individuals with density function $N(a, t)$ [or $M(a, t)$ for the subunits within an individual] equals the integral of the product of age-specific per unit mass growth rates ($g(a)$), average per capita mass ($M(a, t)$) and numbers ($N(a, t)$).

$$D_G(a) \approx \int_0^\infty g(a)m(a, t)N(a, t)\, da = \int_0^\infty g(a)M(a, t)\, da \quad [8]$$

$g(a)$ could also be formulated per capita eliminating the need to include average age-specific mass $m(a, t)$.

Of course, the (maximum growth rate of the subunit populations within individuals may be estimated separately, or the growth rates may be assumed proportional to the growth rate of the parent organism (*allometric growth*). For example, if we assume allometric growth, of say plant subunits, to the maximum total leaf growth rate G_L, the proportionality constants for root, stem and reserves (v_i, $i = r$, s, res) may be estimated from data. Under nonlimiting conditions, the total per plant demand for resources corrected for egestion, respiration and tissue conversion costs (see [4]) is

$$D(a) \approx \frac{G_L(a)(1 + v_r + v_{\text{res}})/\lambda + z(\cdot)M^*(a)}{\beta} \quad [9]$$

where $M^*(a)$ is the mass of the respiring tissues. Note that the egestion term (β) is included in the model to remind us that models of carnivorous plants may need this term. In essence [4] and [9] are equivalent.

Demands for Respiration [z($\cdot$)]. The Q_{10} rule [10] for poikilotherms states that there is a doubling of the base respiration rate (z_0) with every 10°C increase in temperature τ above a threshold θ (see Chapter 2). As applied to our model, the respiration rate is assumed to increase with temperature and we need to estimate this for each species we wish to model.

$$z(\cdot, t, \tau) = z_0[2^{(t - \theta)/10}] \quad \text{for} \quad (\theta \ge \tau) \quad [10]$$

Peters (1983) gives estimates for respiration rates for different sized animals, and as we shall see provides rough useful rules of thumb. Estimating (z_0) per unit mass may be quite difficult for some organisms. An approach might be to estimate $z(\cdot)$ at a specific temperature using a respirometer (Llewellyn 1988), and then using this to solve for z_0 in [10]. A satisfactory estimate for z might be obtained from laboratory or field data given reasonable values for the other variables by solving the metabolic pool model (Gutierrez et al. 1981a) as follows.

$$z(\cdot, t, \tau) = \frac{S(a, t, \tau)\beta\lambda - A(a, \tau)}{M^*(a, \tau)} \qquad [11]$$

Hunger, and hence searching rates (e.g., etiolation in plants), increase as resource levels decline, and respiration rates increase accordingly (see Gutierrez et al. 1981). There are more sophisticated procedures for estimating z_0 and $z(\cdot)$, but the intent here is to outline the functional form of the model without arguing important physiological fine points.

Conversion Efficiency (λ). The efficiency ($0 < \lambda < 1$) of conversion of resources acquired from a lower trophic level to a higher one (prey into predator tissues) may be viewed as a constant, but an age-specific function can be accommodated just as easily. The physiological literature often provides reasonable estimates for this parameter for many taxa.

Egestion (β). As with the other parameters, estimates of the proportion of the food egested $(1 - \beta)$ in animals (and carnivorous plants) must be made, but the methods must be tailored to the species.

RESOURCE ALLOCATION

When shortfalls in resource acquisition occur, organisms must allocate resources according to some genetically fixed strategy (i.e., a priority). Our model assumes, in priority order, that of the resource acquired, a proportion is egested, and then costs of respiration and converting prey to self are incurred, and lastly the portion remaining is assimilated with reproduction having a higher priority than growth. Resource allocation occurs as follows:

1. Egestion ($1 - \beta$) and Respiration ($z(\cdot)M$):

If S is the resource acquired at time t, then $S_A = \beta S - z(\cdot)M > 0$ is the maximum quantity that may be assimilated.
However, if $S_A < 0$, reserves may be used

$$\text{Res}(t + \Delta t) = \text{Res}(t) - S_A > 0,$$

but no growth or reproduction occurs. If $\text{Res}(t + \Delta t) \leq 0$, the organism dies.

2. Allocation rate to reproduction (R), growth (G) and reserves (ΔRes)

If $S_A > 0$ and the reproductive demand rate $D_R > 0$, then

$$R = \min[D_R, S_A]$$

and assimilation to somatic growth (A) is

$$A = G + \Delta\text{Res} = S_A - R$$

with

$$\text{Res}(t + \Delta t) = \text{Res}(t) + \Delta\text{Res}$$

Age-specific Allocation to Reproduction. There are several ways one might allocate resources in an age-dependent manner. In this example we allocate to reproductive subunits within a parent organism of age a at time t and also show how the mass of the parent grows in response. [The same methods may be applied to model assimilation in an age-structured population of parents from their birth until death.] In this model, the age (a_R) of subunits is different from the parent (a), and allocation to reserves is ignored for convenience.

We define [$R(a_R, t)$ and $M_R(a_R, t)$] as the number and mass of reproductive subunits respectively of age $0 \leq a_R \leq x$ at time t. The population density of all subunits is

$$R(t) = \sum_{a_R = 0}^{a_R < x} R(a_R, t) \qquad [12a]$$

and their total mass is

$$M_R(t) = \sum_{a_R = 0}^{a_R < x} M_R(a_R, t) \qquad [12b]$$

The average mass of a reproductive subunit of age a_R at time t is

$$\overline{m}(a_R, t) = \frac{M_R(a_R, t)}{R(a_R, t)} \qquad [12c]$$

If $g_R(a_R)$ is the maximum age-specific per capita growth rate of reproductive subunits of age a_R per unit time, then the total demand of all reproductive subunits during time t to $t + \Delta t$ is D_R.

$$D_R(t) = \sum_{a_R=0}^{a_R < x} g_R(a_R) R(a_R, t) \, \Delta t \qquad [13]$$

Furthermore, if we assume S_A is the food assimilated by the parent of mass $M(a, t)$ (see above), then total growth by all subunit $\Delta M_R(t)$ is the same mass accrued by the parent (i.e., $\Delta M(t)$).

$$\Delta M(t) = \Delta M_R(t) = \min [D_R(t), S_A(a, t)] \qquad [14]$$

The weight gain by reproductive units of age a_R is

$$\Delta M_R(a_R + \Delta a_R, t + \Delta t) = [\phi_R(t) g(a_R) R(a_R, t) \, \Delta t] \qquad [15]$$

The supply/demand ratio

$$0 \le \phi_R(t) = S_A(a, t)/D_R(a, t) < 1 \qquad [16a]$$

scales the potential growth rates to the realized ones, and reduces the maximum number of new reproductive subunits $R^*(a)$ initiated.

$$R(a_R = 0, t) = \phi_R(t) R^*(a) \qquad [16b]$$

In addition it may scale their maximum average initial mass $\overline{m}_0$.

$$M_R(a_R = 0, t) = \phi_R(t) \overline{m}_0 R^*(a) \qquad [16]$$

Last, the weight of the parent containing the developing subunits is

$$M(a + \Delta a, t + \Delta t) = M(a, t) + \Delta M_R(t) - \Delta M_R(a_R = x, t) \qquad [17]$$

The mass of all new born at birth age $a_R = x$ entering the parent population is $\Delta M_R(x, t) = \overline{m}_R(x, t) R(x, t)$. This term is the product of the number of subunits of age $(R(x, t))$ times their average mass $\overline{m}_R(x, t)$ [see 12c].

These computations show how time-varying resource acquisition and allocation affect time-varying size and birth rates. Or course, these birth and growth calculations may be simplified by making assumptions (e.g., the average mass of a newborn is proportional to the average mass of the mothers, Ward et al. 1983a,b), but important information may be lost. These age-specific computations will be reexamined when we address age-structured population models.

The demographic effects of poor nutrition due to low-resource base or unfavorable weather are manifested in small adult size, low-birth rates and weights, and slowed growth rates. These shortfalls may greatly affect competitiveness of the individual in the current generation (Huston 1988), and the effect may have a greater negative impact in subsequent generations via carryover effects on size, vigor, sex ratios, and sundry other factors. We shall show how these factors are included in the model.

THE EFFECTS OF TEMPERATURE

Temperature is not a resource for poikilothermic organisms, but it is a major determinant of the resource demand rate (Sharpe and DeMichele 1977) and of the realized assimilation rate. Both these factors affect many aspects of the biology of a species. The effects of temperature on blue alfalfa aphid developmental rates, fecundity (and size), and survivorship, respectively, were illustrated in Chapter 4, Figure 4.6a–c (cf. Summers et al. 1984). The derivative of the developmental rate function (the acceleration, Figure 4.6d) is parabolic with respect to temperature, and the same shape emerged for gross reproduction and mean survivorship (Fig. 4.6e and f). Normalized, these functions become indices of the favorability of temperature. The shape of the functions is due to the interplay between resource supply and demand, and their possible origins are outlined below.

In the ladybird beetle, *Hippodamia convergens* G-M, the amount of aphid prey eaten per day increases with temperature to a maximum of approximately 25% of body weight at about 12°C (Fig. 6.7a). In contrast, the rate of movement, which is a surrogate measure of respiration costs, increases linearly with temperature (Fig. 6.7b). The limits to food consumption are gut size and/or

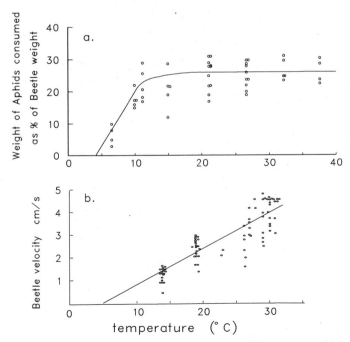

Figure 6.7. *The effects of temperature on ladybeetle* Hippodamia convergens *G-M (a) consumption in the presence of abundant pea aphid prey and (b) on its rate of movement. [Gossard et al. 1976.]*

the rate the food can be processed (see Mills 1981, p. 311), but metabolic costs may continue to grow. Generally, $0 \leq S < D$ and if resources are low and/or the respiration rate is high because of high temperature, D may be much greater than S. Similar arguments can be made for photosynthetic and respiration rates in plants as temperature increases. Of course, the demand rate of an organism varies with age, gender, size, temperature, and other factors. Reproductive demands may be small relative to other demands in some species and may be ignored (e.g., cassava and some parasitoids), or they may be large and must be taken into account (e.g., aphids, bean, cotton, grape, rice, and tomato).

The combined effects of temperature (τ) and resource level on the assimilation rate (A) and the metabolic compensation point in *H. convergens* were modeled using [1], and the results illustrated in Fig. 6.8a (Gutierrez et al. 1981). This model predicts that the optimum temperature for maximum assimilation is approximately 12°C, and that above 36°C respiration demands exceed

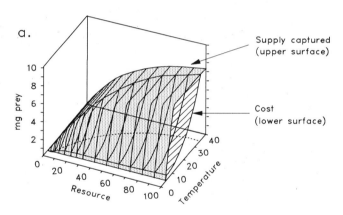

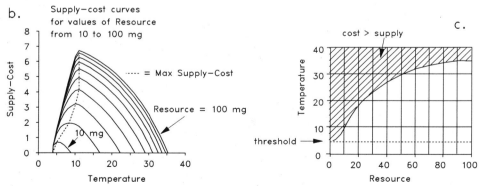

Figure 6.8. *The effects of prey availability and temperature on adult female* H. convergens *G-M: (a) resource acquisition and metabolic costs, (b) the difference between resource acquisition and metabolic costs, and (c) a map of the metabolic compensation points.*

the animal's ability to ingest and process food (Fig. 6.8b). The upper thermal compensation threshold declines with decreasing prey availability (Fig. 6.8c). Similar figures can be drawn for any poikilothermic organism (see Gutierrez 1992).

THE EFFECTS OF THE SUPPLY TO DEMAND RATIO

Growth and Development. In our metabolic pool model, all rates are controlled by the appropriate supply/demand ratio, and hence by factors that affect either the *supply side* (*S*) or *demand side* (*D*) of the ratio. Shortfalls may greatly affect the dynamics of growth and phenology as illustrated by the stylized interplay of supply and demand in cotton (Fig. 6.9; cf. Wang et al. 1977). *Physiological stress* (i.e., $D \gg S$) is most likely to occur when reproductive subunits begin rapid growth (Fig. 6.9a). Under optimal conditions this normally

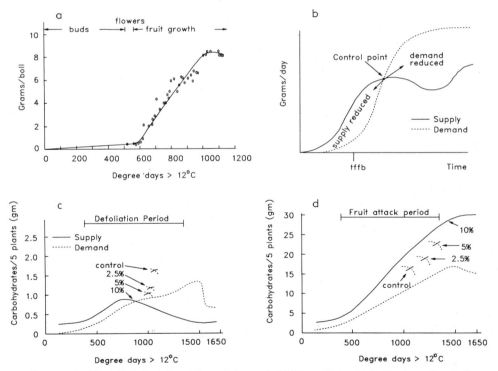

Figure 6.9. *The interaction of supply and demand: (a) The pattern of growth of a cotton fruit, (b) the hypothetical pattern of supply and demand in a cotton crop under near optimal conditions, (c) the effects of different levels of supply side pests on the interaction of supply and demand, and (d) the effects of different levels of demand side pests on the interaction of supply and demand. (Note that the control point occurs when the demand equals the supply under optimal conditions.) [Modified from Wang et al. 1977.]*

occurs because the combined metabolic, reproductive and growth demands exceed the ability of the organism to meet them. This stress is indicated by the control point in Fig. 6.9b. Biotic and abiotic factors that affect the supply side (Fig. 6.9c) cause stunting of the plant and the time of stress to occur earlier. Factors that reduce demands (Fig. 6.9d) cause the stress to be delayed and induce vegetative growth. Abiotic and biotic factors causing reductions in the supply side in plants are poor resource availability (cloudy weather, water and nitrogen stress, etc.), disease, and reductions in root and leaf area due to herbivory. These cause stress and general unthriftiness of the plant. The demand side may be increased by increased demands of fruits and/or the accumulation of nonphotosynthesizing tissues (stem and root). Of course, plants with small or slow growing reproductive units experience stress later (or not at all) than those with large demands.

The applicability of this supply–demand paradigm to animals has been demonstrated (Gutierrez and Wang 1976; Gutierrez and Baumgärtner 1984a,b; Gutierrez 1992). For example, in viviparous aphids, stress may occur when the demand of growing embryos within a female becomes large. In animals, prey may be in short supply and/or be of poor nutritional quality, or of the wrong size or stage. Such supply side considerations are commonly accepted in animal husbandry where feed is allocated to maximize some attribute of production (e.g., meat or eggs). Factors that reduce the demand side in animals are not as obvious as in plants, but low temperature, small size, and quiescence are examples.

In some organisms, growth demands of reproductive subunits may be small relative to overall demands, and may be relatively unimportant (see Gutierrez et al. 1987).

Effects on Survivorship. Common ecological wisdom assumes that the capacity of most species to reproduce exceeds that required for replacement. In the absence of effective natural enemies, populations grow until resources become limiting, which causes fecundity and survivorship to decline. In many species, reproductive capacity may be regulated by resource shortfalls as indicated above (i.e., by aborting embryos and buds that have little chance of completing development and by the reduction in size of surviving individuals). In this strategy, the reproductive units aborted are the ones in which the least amount of energy and the least time has been invested (i.e., minimizing losses, Gutierrez and Wang 1976; Ward et al. 1983a,b). However, if conditions deteriorate further, whole organisms as well as the reproductive subunits they contain may perish.

At the individual level, survivorship (s) is 0 or 1. However, recall that the effects of temperature on blue aphid average longevity gave a parabolic function (see Fig. 4.6c). This same function emerges from the ratio of the realized rate of assimilation at a given level of resource across temperatures to the theoretical maximum at the optimum temperature (τ_{opt}), and is a measure of departure from the optimum.

$$0 \leq l_x = \frac{S(a, \tau) - z(\cdot, \tau)M(a)}{S_{\max} - z(\cdot, \tau_{\text{opt}})M(a)} < 1. \qquad [17]$$

We can view l_x as an intrinsic survivorship function (that includes emigration) under a given set of conditions of temperature and level of food. These effects on survivorship are easily accounted for under variable conditions making estimates of mortality due to natural enemies easier because the background intrinsic rate is already in l_x (e.g., recall the concepts of potential and observed population growth rates $e^{\wp} - e^{\lambda}$, in Hughes 1963, Chapter 4).

REFLECTIONS

Evolutionary theory suggests that reproduction should have a higher priority than growth. This makes sense since making investments in growth is merely one of the costs of progeny production (i.e., Gutierrez and Regev 1983). *To paraphrase an old idea, energy expended in growth is the genome's investment in reproducing itself.* As a caveat, however, there are likely trade-offs between growth, survival, and reproduction, which may depend on environmental predictability (Gatto et al. 1989; Gilbert 1990), hence it is conceivable that in some organisms growth may have a higher priority, but this is not considered here.

Unfortunately, dry matter allocation studies by themselves . . . *shed little light on one of the central problems in the evolutionary approach to physiological ecology: why are resources partitioned exactly as they are* (Pianka 1981). This is obviously a paraphrase of Fisher's rhetorical question, which introduced this chapter. The most common approach used for examining ecological energetics is to develop models that account for the flow dynamics of energy (or mass) through the ecosystems (Odum 1971). This approach provides good estimates of flow rates, but does not help answer Fisher's question. In my opinion, the best approach is via an analysis of per capita energy partitioning considered in an ecological and population context. Energy acquisition and allocation provides a better currency for intra- and intertrophic dynamics as well as a realistic basis for developing multitrophic population models. The development of these models is accomplished by extending the notions of the metabolic pool model to the population level using per capita considerations.

Appendix 6.1.

Effects of Other Essential Resources on Resource Acquisition

Shortfalls in the acquisition rate of one or more essential resources may have important effects on growth and reproduction rates (assimilation), and hence on the dynamics of populations. In plants, net relative growth rates are characteristically parabolic functions of increasing levels of temperature and resources (nitrogen and other inorganic resources) (Shelford's law of tolerance 1931). Similarly, the acquisition of different kinds of nutrients or requisites (e.g., nesting sites, and mates) by animals may affect individual and population growth rates. These functions are the acceleration rates of assimilation at different levels of the factor, and imply that below the peak level slowing of growth (or reproductive) rate is due to shortfall of the resource (supply), and above the level there is an increasingly toxic effect (see Fitzpatrick and Nix 1970). The effects of nutrient levels have been documented for animals in population ecology (Sharpe and Hu 1980; McNeill and Southwood 1978; Stone and Gutierrez 1986; Wermelinger et al. 1987, 1991b; Gutierrez et al. 1994), hence the innovation here is to show how this biology is incorporated in the metabolic pool model for plants, but note that the same model applies for multiple resource acquisition in animals—only the resources sought change.

Fitzpatrick and Nix (1970) viewed these normalized parabolic response functions as indices of favorability for different levels of each factor on the growth rate of temperate pasture plants. They used the product of these indices to assess the combined effect of all factors. Gutierrez et al. (1974b) used the same approach to characterize the limits for aphid population growth. The concave supply demand indices (S/D) estimated from [3] may be used to compute the effects of shortfalls, but not excess of different resources (Gutierrez et al. 1974b, 1988b, 1994). For example, the ability of plants to acquire nitrogen and water from the root zone under varying conditions (Gutierrez et al. 1988b) may affect growth rates.

The per unit mass acquisition model in the text ([3]) is modified for the acquisition of nitrogen (subscript η) and water (ω) by plant roots (Gutierrez et al. 1988a, 1994). The apparency parameter (α_r) now reflects apparency of these two requisites to the root mass searching the soil for them.

$$S_\eta = (1 - e^{-\alpha_r\eta/D_\eta M_1})D_\eta \qquad \text{[A6.1a]}$$

$$S_\omega = (1 - e^{-\alpha_r\omega/D_\omega M_1})D_\omega \qquad \text{[A6.1b]}$$

Nitrogen demand (D_η) may be assumed proportional to the demand for carbon (D_c), and the demand for water (D_ω) is the maximum evapotranspiration rate and may be computed using a biophysical model (e.g., Ritchie 1972). It is useful to cast these functions in the form of concave supply/demand ratios (ϕ_η and ϕ_ω) by dividing both sides of the function by the appropriate demand rate.

$$0 \le \phi_\eta = (S_\eta/D_\eta) < 1 \quad \text{and} \quad 0 \le \phi_\omega = (S_\omega/D_\omega) < 1 \qquad \text{[A6.2]}$$

These ratios are the proportions of the demand for nitrogen and water, respectively, acquired, and these and others shortfalls of essential resources ($i = 1$, ..., n) may be used to scale the potential photosynthetic rate (S_C) to the realized rate (S^*).

$$S^*(\eta, \omega, \cdot) = (1 - e^{-\alpha_1\Gamma/D_C M_1})D_C \cdot \prod_{i=1}^{n} \phi_i \qquad \text{[A6.3]}$$

where Γ is the light energy converted to biomass equivalents. The exponential term in the hunting equation for each resource is the probability of not meeting the demand for that factor. If we divide both sides of [A6.3] by D_C, it becomes the realized carbon supply/demand ratio (ϕ_C). Note that $\prod_{i=1}^{n} \phi_i$ is equivalent to compounding contemporaneous mortality terms (Royama 1981).

In nature, some resources may fluctuate widely over relatively short periods of time (e.g., light and prey), while resources such as soil nitrogen and water for plants and nesting sites, food quality, and other factors for animals may change relatively slowly. Hence, the effects of these factors on growth may be correspondingly fast or slow. In the absence of natural enemies and competition, growth could be limited by the single factor in shortest supply (Leibig's law of the minimum 1840), but realistically it is reduced or possibly limited by the compounded partial shortfalls of all resources [A6.3] (see Gutierrez et al. 1988b; Berryman 1992). Hence, the realized assimilation rate is

$$A(t) = G(t) + R(t) = \{\beta S^*(t, \cdot) - z(\cdot, \tau)M(t)\}\lambda \qquad \text{[A6.4]}$$

Modeling: A Preview

It is generally accepted that the first steps in developing a population dynamics model are to determine the objectives (*strategic versus tactical*, Conway 1984) and outline the nature of the biological questions one wishes to explore. Only then can one select the appropriate mathematical form of the model to satisfy these sometimes conflicting goals. This is not an easy matter, the richness of the biology of species interactions often clouds our ability to formulate models for the dynamics of even supposedly simple systems. The mathematical form of a model is an important consideration as it may limit the amount of biology that can be included and our ability to analyze its properties. Models with even limited amounts of biological complexity may not have analytic solutions and may require numerical simulation. Quite often simplifications of the biology are made to satisfy mathematical constraints, and this severely limits the utility of the model for assessing field problems. The following figure below depicts five potential stages in model development in applied population ecology, as well as some of the trade-offs between model simplicity and complexity. The potential range of applications of different levels of models is also depicted and the details are fleshed out in Chapters 7–11.

No model (*by definition*) can capture the full richness of predator–prey interactions, for if it could it would likely be as incomprehensible as nature itself. However, it is well known that models of the same system having quite different mathematical form and biological assumptions can give qualitatively similar comparisons to data, but lead to grossly different interpretations of the interactions. Anderson and May (1980) developed an analytical model that qualitatively reproduced the temporal dynamics of the larch budmoth by assuming an interaction with a polyhedrosis virus disease. Fischlin and Baltensweiler (1979) achieved equally good results using a simulation model that included the effects of changing nutritional status of the larch trees on budmoth survivorship and fecundity but excluded the virus. Was the increase in virus infected larch budmoth enhanced by poor nutrition, or did the epizootic operate in the absence of food quality? It is known that viral epizootics are very infrequent in tortricids such as the larch budmoth, and are thought to occur in stressed populations. The larch budmoth is not an exception among lepidopterous herbivores (N. Mills, personal communication), and it is safe to say, most of the information on insect viral epizootics in nature is at best anecdotal. The problem of model type, the biology to be included, and the interpretation of model results will hopefully become clear as we explore attempts to model Nicholson's not so simple laboratory data on the population dynamics of the sheep blowfly.

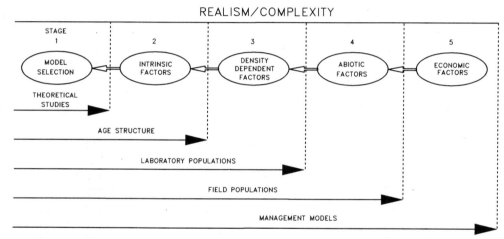

Stages in developing realistic models for use in applied population ecology [cf. Gutierrez and Curry 1989].

7

Simple Single Species Models

Many population models describe nothing known to God or man.
—Gilbert et al. 1976

How the population dynamics of a species can be modeled realistically remains a major issue among ecologists, but considerable progress has been made in the last two decades. The models of Lotka and Volterra, and Gause (1934), Nicholson (1932), Nicholson and Bailey (1935), and later McArthur (1955), May (1973, 1981), and Hassell (1978) lay much of the ground work for modern theoretical population ecology applicable to field problems. Kingland (1985) gives an account of these historical developments. The concepts of functional and numerical response (i.e., *resource acquisition and allocation*) were introduced in Chapters 5 and 6 and this knowledge is assumed here. In this chapter, the dynamics of single species are emphasized, and the underlying assumptions of some commonly used simple population dynamics models are examined and used to lay the underpinnings for the development of models for field applications. In a step-by-step manner, the realism of the metaphysiology outlined in Chapters 5 and 6 is added in Chapters 7–10 culminating in the development of field models. The mathematics of the dynamics models used are discussed in Appendix III. Also reviewed are the differences between difference and differential equations models and, as is convenient, we shall switch between the two forms. Nicholson's (1935) laboratory population experiments on blowfly (*Lucillia cuprina* Wiedman) are used as a common foil here and in Chapter 8 to demonstrate the applicability of different kinds of models. Let us start simply by showing origins and adding the biology implied in Figure 7.1.

a No age structure with density dependent interactions

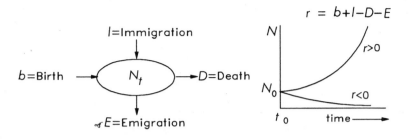

b Plus carrying capacity

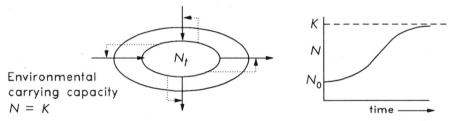

Figure 7.1. *Model with increasing degrees of complexity: (a) simple model and (b) density dependent feedback.*

LOTKA–VOLTERRA ORIGINS

In general, the Lotka–Volterra models yield academically interesting theoretical predictions concerning species interactions, but unfortunately, they are not amenable to field testing, and as such have had little impact in applied ecology. The models simply leave out too many of the biological details. The models used by many applied ecologists, however, have their historical and theoretical roots in these dynamics models and are the starting point for our model.

Simple models often ignore age-structure effects and the biology is condensed into a few parameters. The simplest population (N) model is the Malthusian exponential growth model, since it ignores all constraints on the constant per capita population growth rate (r)

$$\frac{dN}{dt} = rN \qquad [1a]$$

Its solution is found by first separating variables

$$\frac{dN}{N} = r \, dt \qquad \text{[1b]}$$

and integrating,

$$\int \frac{dN}{N} = \int r \, dt \qquad \text{[1c]}$$

yielding

$$\ln N + c = rt \qquad \text{[1d]}$$

If we take the antilogarithm of [1d] and recognize that e^{-c} is a constant, say N_o, we get

$$N_t = e^{-c} e^{rt} = N_o e^{rt} \qquad \text{[1e]}$$

Thus, if we know initial condition N_o, we can predict N_t at time t. This model predicts an ever expanding population if $r > 0$ and a decreasing population if $r < 0$ (see Fig. 7.1a).

Pearl and Reed (1920) and Verhulst (1838) before them recognized that such growth could not continue unabated, and proposed the logistic equation to describe departures from unrestricted growth.

$$N_t = \frac{K}{1 + e^{a-rt}} \qquad \text{[2]}$$

In this model, K is the upper limit to population growth, r is a constant estimating the unrestricted growth rate, and a is a constant of integration. During the 1920's and 1930's there was argument as to whether this model was a *law of population growth*, but this controversy soon died away as the model was tested against data and common sense prevailed.

Lotka (1925) was strongly influenced by Pearl, and this background is evident in the formulation of his general kinetics model [3] for evolving systems (e.g., population growth).

$$\frac{dN}{dt} = F(N) \qquad \text{[3]}$$

Lotka sought to lay a general framework for population dynamics theory that could be used to examine the equilibrium properties of interacting species. His model assumes population growth under constant conditions where the growth rate is a function of density (i.e., the dashed lines in Fig. 7.1b).

A population is thought to be at equilibrium when $F(N) = 0$, where $N = N^*$ is the equilibrium level of the population (see Appendix II). To examine the properties of this general system, he first approximated $F(N)$ near zero using a Taylor's polynomial expansion, [4].

$$\frac{dN}{dt} \approx c_o + aN + bN^2 + cN^3 + \cdots \qquad [4]$$

To satisfy the property $F(N) = 0$, Lotka ignored the constant c_o and arbitrarily all terms of power greater than 2. (The analysis can also be done in per capita form.) We assume for convenience.

$$\frac{dN}{dt} = F(N) = aN + bN^2 \qquad [5]$$

or in per capita form

$$\frac{dN}{Ndt} = (a + bN)$$

Common sense dictates that the per capita growth rate decreases with density and must be positive at low density. If a is the per capita population growth rate, then b must be negative and is the self-limiting effect of population growth. Equation [5] has two roots $N = 0$ and $N = -a/b$ that satisfy $F(N) = 0$.

To integrate [5], we again separate the variables

$$dt = \frac{dN}{\{a - bN\}N} \qquad [5a]$$

and rearrange terms

$$a \, dt - bN \, dt = \frac{dN}{N} \qquad [5b]$$

Substituting [5a] for dt in $bNdt$ in [5b], simplifying, and integrating yields

$$\int \frac{dN}{N} + \int \frac{b \, dN}{a - bN} = \int a \, dt \qquad [5c]$$

(From the chain rule we note that $b/(a - bN)$ in [5c] is the derivative of $-\ln(a - bN)$ with respects to N.) Integrating and combining the constants of integration into c, [5c] yields

$$\ln N - \ln(a - bN) + c = at \qquad [5d]$$

Rearranging and taking the antilogarithm yields

$$\frac{N}{(a - bN)} = e^{-c}e^{at} = Ce^{at} \qquad [5e]$$

It now takes a bit of manipulation to get [5e] into a form recognizable as [2]. If we rearrange terms as follows:

$$N = (a - bN)Ce^{at} = aCe^{at} - bNCe^{at} \qquad [5f]$$

$$(1 + bCe^{at})N = aCe^{at}$$

$$N = \frac{aCe^{at}}{1 + bCe^{at}}$$

and by dividing the top and bottom of the right-hand side by bCe^{at} we get

$$N = \frac{a/b}{1 + \dfrac{1}{bCe^{at}}} \qquad [5g]$$

If $bC = e^d$, then [5b] equals

$$N = \frac{a/b}{1 + e^{d - at}} \qquad [5h]$$

Hence, the limit as $t \to \infty$ is $a/b = K$, and if $r = a$, we get [2]

$$N = \frac{K}{1 + e^{c - rt}} \qquad [5i]$$

Dublin and Lotka (1925) later derived the differential form of [5] written in its more familiar form [6] by considering the effects of age-structure on r

$$\frac{dN(t)}{dt} = rN\left(1 - \frac{N}{K}\right) \qquad [6]$$

where r and K are as defined above. Note that as $N \to K$, $dN/dt \to 0$. Self limiting effects of population growth via competition for resources will be an important theme in our models. (Note also that if $r = a$ and $r/k = b$, we again have the approximation of $F(N)$ in [5].)

APPLICATIONS OF THE LOGISTIC MODEL TO NICHOLSON'S BLOWFLY DATA

Lotka realized that many other biotic and abiotic factors not included in the logistic model affect population growth, yet despite this, numerous attempts have been made to apply variants of the logistic growth model to laboratory populations. May (1973) showed that the oscillations in Nicholson's laboratory populations of blowfly could be adequately modeled using a time delay logistic model.

$$\frac{dN(t)}{dt} = rN(t)\left[1 - \frac{N(t - \tau)}{K}\right] \qquad [7]$$

In Nicholson's experiment, blowfly larvae were fed liver and the adults were fed liver and sugar water. In some studies, the adults were fed food ad lib and the larval food was a restricted but constant daily supply. In other experiments the larvae received food ad lib and the adult food was restricted. We shall return several times to these data.

The time delay (τ) of 9 days used by May differed from the 13–15 day egg to adult period observed by Nicholson (Readshaw and Cuff 1980), and slightly different values of r and very different values of K were required to fit the data (Fig. 7.2). Some ecologists have claimed that this demonstrated the importance of time delays in biological systems, but I view it as a curve-fitting exercise. Other groups have attempted to model Nicholson's data, foremost among them are Gurney et al. (1980), Readshaw and Cuff (1980), and Gutierrez (1992).

OTHER SIMPLE MODELS OF NICHOLSON'S BLOWFLY DYNAMICS

Gurney et al. Model

Gurney et al. (1980) identified the problems with May's model and proposed an alternative model for the class of mechanisms that produce the quasiperiodic fluctuations in Nicholson's data. In the absence of immigration and emigration, they assumed that the population dynamics must be controlled by the net difference between recruitment and death rates to the adults.

$$\frac{dN(t)}{dt} = R(N(t - \tau)) - \mu N(t) \qquad [8a]$$

where μ is the per capita adult death rate, $\tau = 14.8$ days is the egg-to-adult period, and $R[N(t - \tau)]$ below is a time delayed density dependent population birth rate

$$R(N - \tau) = FN(t - \tau)\exp\left[\frac{-N(t - \tau)}{N_o}\right] \qquad [8b]$$

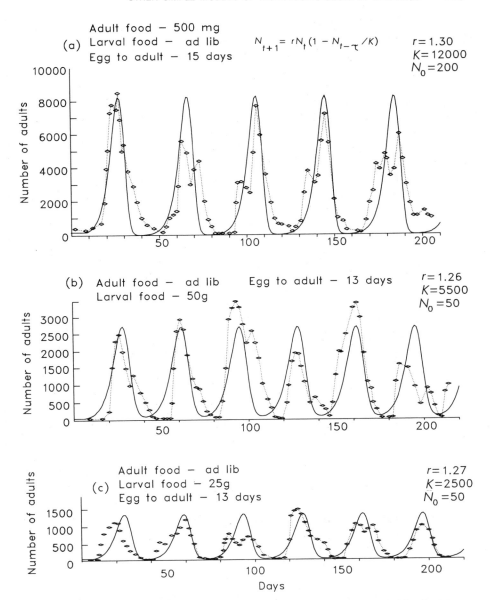

Figure 7.2. Fits of May's delay-logistic model to three sets of Nicholson's blowfly data.

The parameter F is the maximum per capita fecundity and N_o is the population density for maximum reproductive success (i.e., usually near the lowest N). The exponential part is a scalar term for fecundity based on the ratio of the population at time $t - \tau$ (i.e., the eggs that produced the adults at time t) to the density optimal for reproduction. The model assumes that (a) the number of eggs produced per adults depends on current size of the adult population, (b) that all eggs require the same time to become adults (τ), and (c) the probability of an egg maturing to a viable adult depends on the number of competitors of the same age. Gurney et al. wished to determine if the observed fluctuations were . . . self sustaining limit cycles perturbed by experimental uncertainty and demographic stochasticity, or were true quasi-cycles driven by demographic or environmental stochasticity . . . , to obtain via a mathematical analysis of their system . . . *readily extractable dynamics information*, and among other things to explain the *double humps* in the blowfly population data (see Fig. 7.2). The Gurney et al. results were qualitative and are not reproduced here. They concluded, however, that the observed fluctuations arose from a combination of long-developmental delays and the single humped nature of the recruitment curve, and that the double humps occurred when values of N were slightly lower than N_0.

The Readshaw and Cuff Model

Readshaw (1981) strongly objected to the Gurney et al. (1980) model on biological grounds charging that the double humps observed by Nicholson were due to age-dependent charges in fecundity. Using the same data, Readshaw and Cuff (1980) developed a simple difference equation model [9] of adult dynamics based on empirical functions derived from several of Nicholson's experiments

$$N(t + 1) = S_I N(t) + S_F(t - \tau) \, e(t - \tau) N(t - \tau) \qquad [9]$$

where t is time in days, $N(t + 1)$ is the number of adults surviving from the previous day [$S_I N(t)$] plus the surviving newly emerged flies derived from surviving eggs laid τ days earlier [i.e., $S_F(t - \tau) \, e(t - \tau) N(t - \tau)$], $S_I = 0.8$ is an average daily survivorship value estimated from the data, and $\tau = 15$ days is the correct egg-to-adult period. (Note that the time delay implies age-structure.)

The function e is the per capita adult fecundity

$$e(t) = 3.59 \, f_a(t - \tau) - 0.5 \qquad [10]$$

with $f_a(t - \tau) = $ the milligrams liver per adult fly at time $t - \tau$. The function S_F is the effect of prior nutrition on egg-to-adult survivorship

$$S_F(t) = 1 - e^{0.154 - 0.022 f_i(t - \tau)} \qquad [11]$$

where $f_i(t - \tau)$ equals the milligrams liver per egg given at the time $(t - \tau)$ the eggs were laid. Functions $e(t)$ and $S_F(t - \tau)$ are depicted in Fig. 7.3a and b. In their paper, Readshaw and Cuff noted the similarities of their model to those proposed by Maynard Smith (1974) and Taylor and Sokal (1976). The Readshaw–Cuff simulation of Nicholson's data are presented in Fig. 7.4.

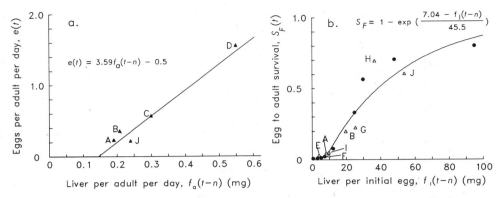

Figure 7.3. *The relationships of blowfly fecundity (a) and survivorship (b) to nutrition. [Redrawn from Readshaw and Cuff 1980.]*

Gutierrez and Baumgärtner Model

Gutierrez (1992) interpreted the Readshaw–Cuff functions $e(t)$ and $S_F(t)$ using the metabolic pool model introduced in Chapter 5. Specifically, he showed that $e(t)$ is predicted by the metabolic pool model ([12], Gutierrez et al. 1981) where the G–B functional type II response model [f (mg liver/egg)] is substituted for f_a in [10]

$$e(t) = \left\{ \beta f \left(\frac{\text{mg liver}}{\text{egg}}, t \right) - r_o(t) \right\} \lambda \qquad [12]$$

The G–B model predicts the per capita acquisition rate of liver by adults, β is the fraction of food not egested, r_0 is the per adult respiration rate, and λ is the efficiency of converting liver to blowfly egg numbers. Under constant conditions, this model is also linear (see Fig. 7.3A, [10]). The constant 0.50 in [10] equals λr_0 in [12] and is the respiration rate in milligrams liver per adult, the metabolic compensation point is 0.14 mg liver adult^{-1} day^{-1} estimated by solving for [10] equal to zero, and $\lambda \beta = 3.59$ eggs produced per milligram of liver per adult per day above the compensation point.

Egg production in ladybird beetles has also been shown to increase linearly with increasing ratios of milligrams prey to milligrams predator (Gutierrez et al. 1981; Gutierrez 1992). Similar functions were obtained by Beddington et al. (1976) for several species.

The egg-to-adult survivorship function [11] may be interpreted in terms of the ratio of the initial per capita supply of liver per egg to the per capita demand required to produce a mature larva of maximum size (i.e., supply/demand = mg liver/larval demand, Gutierrez et al. 1981). The concave shape of [11] is explained by noting that as the per capita consumption of liver consumed increases, survivorship increases but at a decreasing rate (see Baumgärtner et al. 1981; i.e., *diminishing returns*). At very low levels of food, some larvae

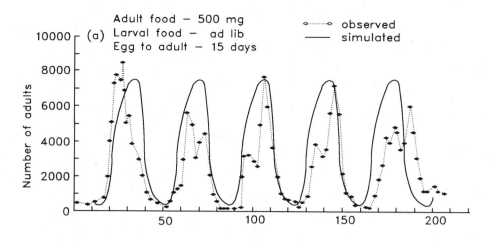

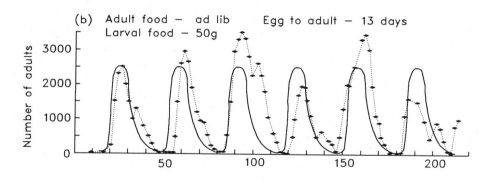

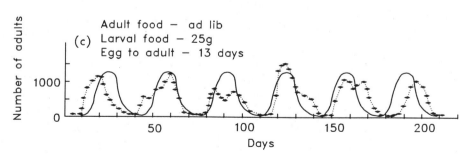

Figure 7.4. Simulation of three sets of Nicholson's blowfly data using the Readshaw and Cuff model (1980), with the figure redrawn. [Reprinted with permission from Readshaw and Cuff 1980.]

get sufficient quantities to produce small but viable pupae, but the fraction that do not increases with competition (i.e., Nicholson's *scramble competition*). The important point is that if the food acquired is less than the demand during the immature period, size, fecundity, and survivorship are adversely affected. If the Gutierrez equivalent functions are substituted for $e(t - \tau)$ and $S_F(t - \tau)$ in the Readshaw–Cuff model [9], the same time lag τ would have to be included to produce the lag related oscillations. In Readshaw–Cuff, however, empirical functions are introduced for the numerical response and survivorship, and in the latter a mechanistic model of the process of resource acquisition and allocation would be used. We shall pursue this problem in an age-structure context in Chapter 9.

LINKS OF THE METABOLIC POOL MODEL TO THE LOGISTIC MODEL

The links between the metabolic pool and the logistic models are at best indirect, as we can only demonstrate that both exhibit the property of diminishing returns (Berryman et al. 1995). Using Lotka as our starting point but casting F as the per capita ratio dependent model, where K is the constant resource base for N, we get the general expression

$$\frac{dN}{dt} = F\left(\frac{N}{K}\right)N \tag{13}$$

A Taylor's expansion of $F(N/K)$ yields

$$\frac{dN}{Ndt} \approx a + b\frac{N}{K} + c\left(\frac{N}{K}\right)^2 + \cdots \tag{13a}$$

and if we keep only the first two terms for our linear approximation we get

$$\frac{dN}{dt} \approx aN + \frac{bN^2}{K} \tag{13b}$$

If $a > 0$ and $b < 0$, the model is clearly logistic when expressed in a ratio dependent form.

The metabolic pool model is also ratio dependent (Gutierrez et al. 1994), and if it is substituted for $F(N/K)$ in [13], then

$$\frac{dN}{dt} = \lambda N\left\{\beta F\left(\frac{K}{N}\right) - r_o\right\} \quad \text{and} \quad F\left(\frac{K}{N}\right) = (1 - e^{-\alpha K/DN})D \tag{14a}$$

Note that the form of the G–B model requires that the ratio be K/N. If we simplify by multiplying through by λN, we get

$$\frac{dN}{dt} = aF\left(\frac{K}{N}\right)N - bN \qquad [14b]$$

One can see that [14a,b] have properties of the logistic model because for constant K, $dN/dt \to 0$ as N increases, and of course $dN/dt \to 0$ when $N \to 0$. A plot of dN/dt on N yields the familiar humped function characteristic of the logistic model intercepting the N axis at zero and at a carrying capacity $DN < K$. However, we should note that the interplay of the two components in [13b] and [14b] are different despite the fact that they yield similar net results. The birth rate aN in [13b] is linear, while the analogous term $F(K/N)N$ is [14b] is concave and saturates to DN (see [14a]). The cost term in [13a] is bN^2/K, which is obviously nonlinear and increasing with N, while the analogous term in [14b] is linear in N. The net in both cases is a humped function, that is, different forms but similar results.

Although the evidence is indirect, and despite my earlier criticism of the logistic model as May applied it to Nicholson's blowflies data, we have gone full circle to show some links. This simple metabolic pool model with an appropriate time lag yields predictions similar to May's (1973) model, but they are not reproduced here. We defer applications to Chapter 9 where age and mass structure and the effects of food supply–demand are included.

A METABOLIC POOL MODEL OF THE DYNAMICS OF CLADOCERAN POPULATIONS IN A CLOSED SYSTEM

In another closed system (Fig. 7.5a), Arditi et al. (1991), Arditi and Saïah (1992) and Arditi and Ginsburg (1992) proposed prey- and ratio-dependent models to examine the dynamics of four species of different sized cladocerans having different search behaviors in serially linked small containers. A constant supply of food (yeast) in water entered the first container and the cladoceran population extracted food to grow and reproduce. The remaining food, but not the cladocerans, was allowed to flow into the next container where that population again extracted food with the remaining unused portion allowed to flow to the next container, and so on down the series. The populations were allowed to go to equilibrium, which in many cases was extinction. The dynamics reflected the food available to the population in the container, the searching behavior of the species, their size and physiological attributes, and possibly increasing waste products from higher containers.

The two species that searched for food throughout the container went extinct in all but the first container. The two species that fed on surfaces of the container or on the surface of the water produced population equilibrium levels that decreased with distance from the first container until no populations were supported. The use of the shunt to put nutrient directly into the last container showed that populations could also develop there and obviated waste products

a.

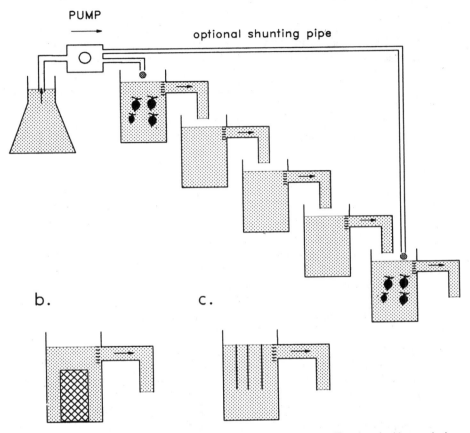

Figure 7.5. *Experimental design of the cascade of five containers [Reprinted with permission from Arditi and Saïah 1992]: (a) alga are pumped at a constant flow rate from a stirred stock container into a cascade each containing an initial inoculum population of cladoceran that is allowed to reach equilibrium but not to migrate between containers. (Note that the optional shunt pipe allows nutrient to go to the first and fifth container to test the hypothesis that it is wastes and not depletion of nutrients that limit population development in the last container.) In some experiments the containers were modified: (b) the dark area is a refuge from predation by free swimming species, and (c) the lines represent strips of mesh that increase the feeding area of species that fed on the side of the container.*

as a limiting factor. Modifications of the environment were made by adding refuges for the food in the case of free swimming species (Fig. 7.5b). This increased the number of vessels that supported equilibrium populations. In contrast, increasing the feeding surface area by the addition of strips of mesh for surface feeding species (Fig. 7.5c) decreased the number of containers reaching positive equilibrium.

Arditi et al. (1991) showed that some dynamics could be explained by a prey-dependent model and others by a ratio dependent model, but not all of the cases could be explained by one model. These data and models also became grist for controversy. For example, Ruxton and Gurney (1992) reinterpreted the data and criticized Arditi et al. (1991) for using a ratio-dependent model in some cases and a prey-dependent model in other cases. Schreiber and Gutierrez (unpublished) showed that dynamics model [15] incorporating the ratio-dependent metabolic pool functional model explained all of the cladoceran experiments including those where the level of food and the environment were altered to increase or reduce refuges for the food against the different species of cladocerans (Arditi et al. 1991, Arditi and Ginsburg 1992).

The generic model for the ith species in the jth container is

$$\frac{dN_{ij}}{dt} = \left\{ \beta_i f\left(\frac{R_{j-1}}{N_{ij}}\right) N_{ij} - r_o N_{ij} \right\} \lambda_i = \beta_i \lambda_i D_i N_{ij}(1 - e^{-\alpha_i R_{j-1}/D_i N_{ij}}) - \lambda_i r_i N_{ij}$$

$$= \theta_i D_i N_{ij}(1 - e^{-\alpha_i R_{j-1}/D_i N_{ij}}) - r_i N_{ij} \qquad [15]$$

where R_0 is the food rate entering the first flask, R_{j-1} is the food resource flowing into the jth flask from the previous flask (i.e., the food not consumed by the $j-1$ population), and α_{ij} is the proportion of the flask that can be searched by the ith species of cladoceran during dt. Free-swimming cladocerans search most of the container and have a high α, while those that feed only on surfaces have a low one, and $1 - \alpha_{ij}$ is the species specific proportion of food not accessible (i.e., in a refuge).

All other factors being equal, the major determinant of whether equilibrium populations developed in the downstream flask was the apparency rate α_i of the ith species. Species with high values deplete the food in the first containers allowing little to flow to subsequent ones. Species with smaller values of α_i allowed populations to develop in more downstream flasks. When refuges were created for the food against free swimming cladocerans having a large α_i, populations developed in more downstream populations. Conversely, the addition of more feeding surfaces for surface feeding species increased food availability (increased α) and resulted in fewer downstream equilibrium populations.

EPILOGUE

Attempts to use simple models to predict the dynamics of even simple systems in nature have met with checkered success, likely because the biology has been overly simplified to make the mathematics tractable. In general, most of these models have used the wrong currency (numbers rather than energy or biomass) and the dynamics have been isolated from the base resource of energy upon which the system depends (i.e., plants, soil nutrients). These models may capture some of the underlying dynamics of a system, but to do this models must include a modicum of realism.

Greater success has accrued with laboratory studies as seen in the various attempts to model the dynamics of Nicholson's blowfly data using simple differential and difference equation models. The models often qualitatively produced the right patterns but lacked explanatory value (May 1973; Gurney et al. 1980). The difference equation model of Readshaw and Cuff (1980) included implied age structure but the birth and death rate functions were derived from the data they sought to reproduce. All of these simple models used time delays as a surrogate for age structure. Despite deficiencies, the Readshaw–Cuff model was the most convincing as the functions they proposed for fecundity and survivorship are predicted by theory and have considerable validation in the literature (see Beddington et al. 1976). The behavior of the metabolic pool model provides some indirect links to logistic theory, and hence to May's time-delay logistic model.

Substitution of the metabolic pool model for the Readshaw–Cuff functions still requires a time lag to reproduce the blowfly dynamics. Finally, a dynamics model that includes the metabolic pool model predicts the complicated dynamics of the four species of cladocerans studied by Arditi et al. (1991) and Arditi and Ginsburg (1992) in all environments and levels of food. These studies show the importance of bottom-up effects on baseline population dynamics.

8

Simple Models of Multitrophic Interactions

Big fleas have little fleas
Upon their backs to bite 'em
And little fleas have lesser fleas
And so at infinitum
—Robert Hegner (1938)

Models of complicated trophic interactions seek, among other things, to help us explain trophic dynamics (May 1973) and the reasons for the persistence of the interacting species. To do the first, we analyze the properties of the model describing the system (i.e., stability analyses, see Appendix II), and ask the question Does the model have a positive equilibrium level(s) and is it stable? The second question is considerably more difficult, but the analysis follows a similar path and asks the question Does the species persist? By stable, we mean that if a system is perturbed from equilibrium (the point of zero population growth), does it return to equilibrium or find a new one (metastability). Unstable systems may be chaotic or may go extinct. Increasingly, more biological detail is being included in simple models (see Abrams 1994) and spatial considerations are becoming more important (Hassell et al. 1991; Murdoch 1994) to help stabilize models and to explain the persistence of species interactions. The need for such spatially distributed meta-population models has yet to be demonstrated, and as we shall see later in this text the idea is not dismissed. This chapter begins along traditional lines and then proposes new models for food chains and food webs.

FOOD CHAINS

Lotka (1925) expanded his general kinetics equation to encompass the competitive interactions of n species

$$\frac{dN}{dt} = F(N_1, N_2, \cdots, N_n, W, Q) \tag{1}$$

In this model, environmental factors (W) are assumed constant and the genetic (Q) parameters are unknown and neglected. When applied to single populations, the model is the logistic model, hence suppose that we have a two-species system composed of competing populations N_1 and N_2. The dynamics of that system might be written as the coupled logistic models widely seen in the literature.

$$\frac{dN_1}{dt} = r_1 N_1 \left(1 - \frac{N_1 + aN_2}{K_1}\right), \tag{2a}$$

$$\frac{dN_2}{dt} = r_2 N_2 \left(1 - \frac{bN_1 + N_2}{K_2}\right), \tag{2b}$$

where a is the per capita effects of N_2 on N_1 and b is the effect of N_1 on N_2, r_1 and r_2 are per capita growth rates, and K_1 and K_2 are environmental carrying capacities of species 1 and 2, respectively. To analyze the behavior of the system, we take Taylor's expansions of [1] with respect to each species separately to approximate the unknown functions $F_n(\cdot)$. The details of this analysis are in Appendix II.

At about the same time the Italian mathematician Volterra (1926, 1931) took up the question of general predator–prey models and proposed the following predator–prey system

$$\frac{dN_1}{dt} = r_1 N_1 - aN_1 N_2 \tag{3a}$$

$$\frac{dN_2}{dt} = hN_1 N_2 - \mu N_2 \tag{3b}$$

where r_1 is the per capita birth rate of N_1 and a is a term relating the predation effects of N_2 on N_1, h relates the conversion of prey attacked to N_2, and μ is the intrinsic death rate of the predator. This model assumes that the same prey death rate may accrue by doubling either N_1 or N_2 (e.g., [3a]). This assumption is obviously weak. Suffice it to say, these models are difficult to relate to the observed dynamics of species in nature, and hence we proceed to the development of realistic models ignoring the complexities of population dispersion and dispersal.

Addition of Realism

As a first step, we incorporate the dynamics of the metabolic pool model introduced in Chapters 5 and 6 to a set of coupled general first-order differential equations often used to describe simple Lotka–Volterra type prey–predator dynamics

$$\frac{dN}{dt} = f(N)N - g(N, P)P \tag{4a}$$

$$\frac{dP}{dt} = hg(N, P)P - \mu P \tag{4b}$$

where $f(N)$ is the per capita birth rate of prey (N), $g(N, P)$ is the predator (P) per capita consumption rate of prey, h converts prey eaten to predator offspring, and μ is the per capita predator death rate (e.g., Arditi and Ginsburg 1989). The biology of trophic interactions as subsumed in f, g, h, and μ in simple dynamics models [4]. The convertibility of mass and number units has been assumed in the literature, and is exploited here. As all organisms are faced with the same problems of resource acquisition and allocation and all are predators in a general sense, then the models of their dynamics should be homogeneous, namely, that the models should have all of the same components. As we shall see, this property is easily met.

Mass ratio-dependent models of form $g(M_i/M_{i+1})$ may be used for both predator (M_{i+1}) and prey (M_i) masses, and the trophic interactions [4] may be rewritten as follows:

$$\frac{dM_i}{dt} = f\left(\frac{M_{i-1}}{M_i}\right) M_i - g\left(\frac{M_i}{M_{i+1}}\right) M_{i+1} \tag{5a}$$

$$\frac{dM_{i+1}}{dt} = hg\left(\frac{M_i}{M_{i+1}}\right) M_{i+1} - \mu M_{i+1} \tag{5b}$$

where f, g, h, and μ in [2] are now in mass units and M_{i-1} is the prey's resource base (Guitierrez and Baumgärtner 1984a,b). This model makes predator fecundity and prey mortality functions of mass acquisition rates, but it does not resolve the issue of time varying changes in the average size (quality) of individuals, the demands of which determine resource acquisition rates and trophic interactions and population dynamics (i.e., in general, large predators require more food than small ones). This suggests that species mass and number dynamics must be linked and the ratio-dependent functional response model must reflect the trophic level size dependence. The linkage between mass and number dynamics is easily accomplished (Gutierrez 1992) as

$$\frac{dN}{dt} = q_i \left[\frac{dM_i}{dt} \right] \qquad [6a]$$

and that for predators is

$$\frac{dP}{dt} = q_{i+1} \left[\frac{dM_{i+1}}{dt} \right] \qquad [6b]$$

where the constants q_i and q_{i+1} convert mass to numbers. This model still assumes constant average size of prey and predator, and this can be resolved only by including the biology of resource acquisition and allocation.

Including the Metabolic Pool Model

Assume we have a tritrophic system composed of a plant (M_1), a herbivore (M_2), and a predator (M_3) written in a per unit mass form (Gutierrez et al. 1994). The mass dynamics for the plant trophic level (M_1) can be written as

$$\frac{dM_1}{dt} = \{\theta_1(1 - e^{-\alpha_1 M_0/D_1 M_1})D_1 - r_1 M_1^{b_1}\}M_1 - (1 - e^{-\alpha_2 M_1/D_2 M_2})D_2 M_2 \quad [7a]$$

where the first term in [7a] is the metabolic pool model (see Chapters 5 and 6 for definitions) incorporating the acquisition function and per unit mass respiration ($r_1 M_1^{b_1}$), which increases with crowding ($M_1^{b_1}$, with $0 \leq b_1 < 1$), and the last term is the rate of herbivory. The number (X_1) dynamics for the plant may be computed as

$$\frac{dX_1}{dt} = \frac{dM_1}{m_1 dt} \qquad [7b]$$

but now $m_1 = \phi_1 m_1^*$ is the average size of an individual obtained as the product of the maximum size (m_1^*) and the overall supply/demand ratio [$\phi_1^* < (1 - e^{-\alpha_1 M_0/D_1 M_1}) = \phi_1$] where $\phi_1^* m_1^*$ is the minimum viable size of an individual (Gutierrez 1992; Chapter 7). Similar mass–number relationships exist at all trophic levels and are not discussed further.

The biomass dynamics of the herbivore (M_2) and the predator (M_3) can be written as

$$\frac{dM_2}{dt} = \{\theta_2(1 - e^{-\alpha_2 M_1/D_2 M_2})D_2 - r_2 M_2^{b_2}\}M_2 - (1 - e^{-\alpha_3 M_2/D_3 M_3})D_3 M_3 \quad [8]$$

and

$$\frac{dM_3}{dt} = \{\theta_3(1 - e^{-\alpha_3 M_2/D_2 M_3})D_3 - r_3 M_3^{b_3}\} M_3. \tag{9}$$

It is quite simple to extended this symmetrical model to multiple prey and/or multiple predators and to include age structure (see Gutierrez et al. 1988a, 1993; Schreiber and Guiterrez 1996). In summary, we note that the trophic parameters θ, D, and α may vary over time, α and D also vary with organism size, D and r vary with temperature and crowding, and the supply/demand ratios (ϕ) are composite functions summarizing resource acquisition success. This complexity is included in age structure in Chapters 9 and 10.

Evaluating Model Realism

Royama (1992) proposed criteria to which all population models must conform. These criteria are normally expressed as per capita growth rates (R-functions, cf. Berryman et al. 1995), but here we use the per unit biomass rates (e.g., for the herbivore $R_2 = dM_2/M_2 dt$). The criteria were originally developed for two trophic levels, and were extended to multiple trophic levels by Berryman et al. (1995). To evaluate the realism of a trophic level model, the population growth rate (say of level 2) must be evaluated as follows: one population (say M_1) is varied and the other two (M_2, M_3) are held constant. These relationships are indicated by the notation $R_2(M_1|M_2, M_3)$.

The criteria for the herbivore level specifically, or by analogy any trophic level, are

1. $R_2(M_1|M_2, M_3)$ is the per unit mass herbivore growth rate, and it should increase at a decreasing rate to a horizontal asymptote when the herbivore and predator biomass are held constant. This occurs because the constant predator density has a decreasing effect on the herbivore that reaches a maximum growth rate as its resource increases.

2. $R_2(M_3|M_1, M_2)$ should be monotonically decreasing with increasing predator biomass when the plant and the herbivore biomass are held constant.

3. $R_2(M_2|M_1, M_3)$ should be parabolic as herbivore biomass increases from zero to the carrying capacity and plant and predator biomass are held constant. This criterion is met only when crowding effects on respiration are such that $b \geq 1$, otherwise the function is sinusoidal with an R_2 intercept and the positive part of the hump is shifted to the right.

4. All trophic equations must contain the same terms (i.e., be functionally homogeneous, Berryman et al. 1995).

Two additional biological criteria were proposed by Gutierrez et al. (1994): namely, that $R_2 < \theta_2 D_2 - r_2 M_2^{b_2}$ because there is a maximum per capita growth rate, and $f(M_1/M_2)M_2 \leq \alpha_2 M_1$ because the predator cannot consume more prey than are available. This trophic dynamics model satisfy all of these criteria.

Analysis of Isoclines and Fixed Points

The properties of this tritrophic model are illustrated using phase-plane analysis in which the population dynamics of any two trophic levels may be represented by their zero growth isoclines (Appendix II; Rosenzweig and MacArthur 1963; Crawley 1992). Normally, the zero-growth isoclines are determined by setting the trophic dynamics model to zero and solving explicitly for one of the adjacent trophic levels. However, unlike the herbivore (M_2) and predator (M_3) isoclines, the plant (M_1) isocline cannot be solved explicitly and is examined analytically.

The plant isocline has two forms that depend on the relationship of α_2 to $\theta_1 D_1$. The plant isocline is parabolic for $\alpha_2 > \theta_1 D_1$ with its peak increasing and shifting to the right as plant resources (M_0) are enriched and declining as herbivore demand (D_2) increases (Fig. 8.1a). The carrying capacity of the environment for the plant population is $M_1 = (\theta_1 D_1/r_1)^{1/b_1}$ which is the ratio of the potential growth rate to the respiration rate raised to the power b^{-1} (i.e., the reciprocal of the degree of self-limitation). The general shape of this isocline has been widely reported (Rosenzweig and MacArthur 1963), but the influence of herbivore demand on the plant isocline had not previously been recognized. When $\alpha_2 < \theta_1 D_1$, the plant isocline has a maximum vertical asymptote at $M_1 = [(\theta_1 D_1 - \alpha_2)/r_1]^{1/b_1}$ and intercepts the M_1 axis at $(\theta_1 D_1/r_1)^{1/b_1}$ (Fig. 8.1b). Arditi and Ginsberg (1989) also found this form of isocline using their ratio-dependent functional response model. The transition between the two isocline forms occurs rapidly near $\alpha_2 = \theta_1 D_1$ (see Gutierrez et al. 1994).

Solving the herbivore model [8] explicitly for M_1 we get the herbivore isocline:

$$
\begin{aligned}
M_1 &= \frac{-D_2 M_2}{\alpha_2} \log_e \frac{\theta_2 D_2 M_2 - r_2 M_2^{1+b_2} - D_3(1 - e^{-\alpha_3 M_2/D_3 M_3}) M_3}{\theta_2 D_2 M_2} \\
&= \frac{-D_2 M_2}{\alpha_2} \log_e \frac{\text{realized herbivore population growth rate}}{\text{potential herbivore population growth rate}}
\end{aligned}
\tag{10}
$$

This isocline also has two cases that depend on whether the inequality $\alpha_3 > \theta_2 D_2$ is met. When the inequality is met, a series of C-shaped herbivore isoclines (see Metzgar and Boyd 1988; Fig. 8.1c) are obtained as predator biomass (M_3) increases, and when $\alpha_3 < \theta_2 D_2$ the isocline is sigmoidal and rises to an asymptote (Fig. 8.1d). Both isoclines have the same maximum carrying capacity $(\theta_2 D_2/r_2)^{1/b_2}$. The herbivore model ($M_2$) cannot be solved explicitly with respect to M_3, but we see that it has the same form as the M_1 isocline for fix-plant resources, and hence has the same attributes. The isoclines of all intermediate trophic levels depend on similar criteria.

a. Plant isocline for $\alpha_2 > \theta_1 D_1$

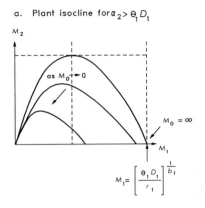

b. Plant isocline for $\alpha_2 < \theta_1 D_1$

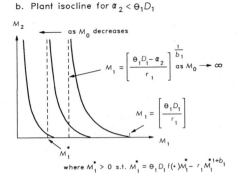

c. Herbivore isocline for $\alpha_3 > \theta_2 D_2$

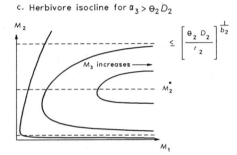

d. Herbivore isocline for $\alpha_3 < \theta_2 D_2$

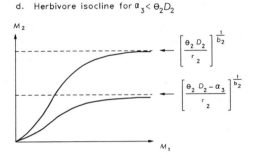

Figure 8.1. *Possible isoclines for two trophic levels in a tritrophic system. The form of the isoclines for the plant and herbivore levels (i.e., all intermediate trophic levels) depends on the inequalities between the apparency of the trophic level to its consumer and its potential per unit growth rate (e.g., $\alpha_{i+1} > \theta_i D_i$ and $\alpha_{i+2} > \theta_{i+1} D_{i+1}$, respectively, see text for definitions): (a) the plant isocline is parabolic when $\alpha_2 > \theta_1 D_1$ with the height and width varying with resource level, (b) the plant isocline has a vertical asymptote when $\alpha_2 < \theta_1 D_1$, (c) the herbivore isocline is C-shaped when $\alpha_3 > \theta_2 D_2$ and its upper and lower asymptotes converge and move to the right with increasing predator numbers, and (d) the herbivore isocline has a horizontal asymptote when $\alpha_3 < \theta_2 D_2$. [Reprinted with permission from Gutierrez et al. 1994.]*

For the predator isocline, we get

$$M_2 = \frac{-D_3 M_3}{\alpha_3} \log_e \frac{\theta_3 D_3 - r_3 M_3^{b_3}}{\theta_3 D_3}$$

$$= \frac{-D_3 M_3}{\alpha_3} \log_e \frac{\text{realized per unit predator growth rate}}{\text{potential per unit predator growth rate}}$$

[11]

which is an increasing function with an asymptote at the maximum carrying capacity $M_3 = (\theta_3 D_3 / r_3)^{1/b_3}$.

In summary, the herbivore (M_2) and predator (M_3) isocline can be solved explicitly, but the plant (M_1) isocline must be evaluated numerically. The maximum carrying capacity of all trophic levels is defined by energetic constraints [i.e., $M_i = (\theta_i D_i / r_i)^{1/b_i}$]. The top trophic level isocline is always asymptotic, while lower trophic level isoclines have two forms that depend on the specific inequality (e.g., $\alpha_{i+1} > \theta_i D_i$). Based on these inequalities, four pairwise combinations of (M_1, M_2) isoclines are possible and represent different trophic relationships. Figure 8.2 shows the effects of increasing or decreasing the level of each factor on trophic isoclines. Figure 8.3 shows possible equilibrium dynamics of the four tritrophic interactions. We caution that simply having the equilibrium point to the left of the peak is not sufficient to assure instability (see May 1973).

Trophic parameters may vary considerably across taxa, and if we wish to examine the qualitative dynamics of a specific system we must estimate them for each trophic level (e.g., Fig. 8.2). However, useful estimates of the parameters can be obtained from the literature and argued from intuition. For example, Peters (1983) summarizes several relationships between physiological processes and body mass that are useful in the context of the model. The demand rate D (ignoring the subscript) can be qualitatively estimated from the relationship between the growth rate G and average body mass m

$$G = am^{3/4} \qquad [12]$$

where a is a constant. We can get a reasonable estimate of the per unit growth demand rate (D_g) by dividing both sides of [12] by m (see Gutierrez et al. 1994).

$$D_g = \frac{G}{m} = am^{-1/4} \qquad [13a]$$

This inverse relationship [13a] can then be used to estimate the per unit demand rate D corrected for all metabolic costs and egestions (see Chapters 5 and 6).

$$D = \frac{(D_g + rM^b)}{\theta} = \frac{(am^{-1/4} + rM^b)}{\theta} \qquad [13b]$$

where rM^b is the per unit respiration rate that includes the effects of crowding and body size.

Using this inverse relationship of size to demand and our intuition about apparency of lower trophic levels to higher ones as guides, we can formulate rough estimates of the inequalities for trophic interactions, characterize the isoclines, and examine their qualititative dynamical properties. Some of the possible interactions are explored below (cf. Guiterrez et al. 1994).

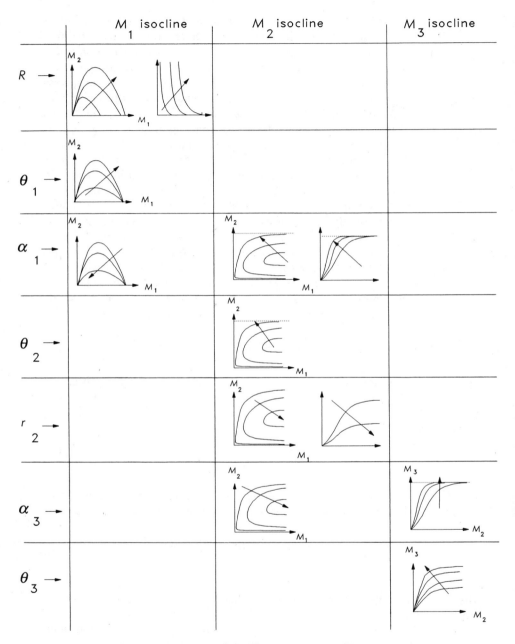

Figure 8.2. *Some effects of trophic level parameters on isocline behavior.*

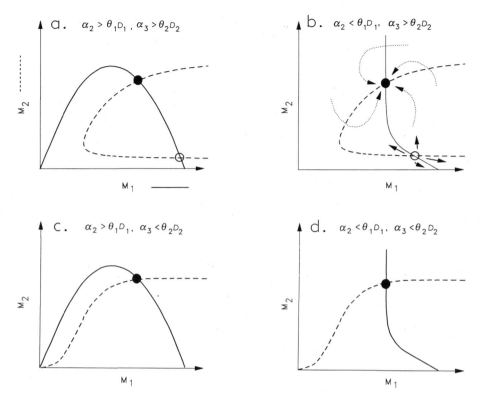

Figure 8.3. *Combinations of isoclines based on trophic inequalities (stable (●) and unstable (○) equilibrium points). [Reprinted with permission from Gutierrez et al. 1994.]*

Case Studies

Case 1: ($\alpha_2 > \theta_1 D_1$ *and* $\alpha_3 > \theta_2 D_2$). The tritrophic dynamics depicted in Figure 8.4a represent the interaction of an efficient predator and an efficient herbivore as might be expected in the case of the successful biological control of an herbivore. Note that the cross sections at equilibrium are given as inserts in the figure. The isocline inequalities are $\alpha_2 > \theta_1 D_1$ and $\alpha_3 > \theta_2 D_2$, suggesting a parabolic shaped isocline for the weed, a C-shaped isocline for the herbivore, and an asymptotic one for the predator. For biological control to occur at high-plant density, the equilibrium (point *a*) must be to the far right of the hump of the plant isocline. Note, however, that the second tritrophic equilibrium (point *b*) is a saddle point, and that population densities of plant and herbivore below this lead to the extinction of the herbivore and of the predator. The qualitative predictions of this model compared well to the observed dynamics of the biological control of various pest species (see Gutierrez et al. 1994).

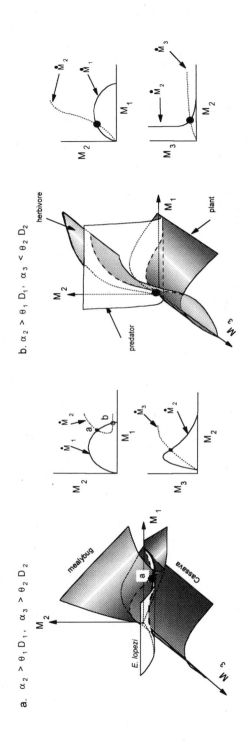

a. $\alpha_2 > \theta_1 D_1$, $\alpha_3 > \theta_2 D_2$

b. $\alpha_2 > \theta_1 D_1$, $\alpha_3 < \theta_2 D_2$

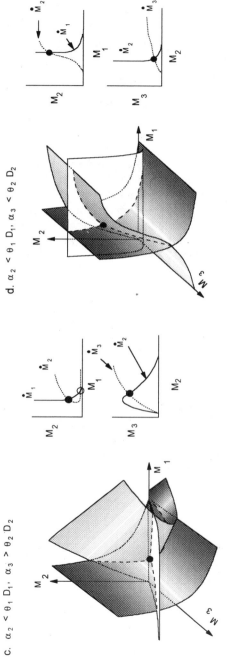

Figure 8.4. Tritrophic analysis of isosurfaces for some tritrophic systems consisting of a plant, a herbivore, and a natural enemy for the following inequalities (see text for discussion): (a) biological control of an herbivore ($\alpha_2 > \theta_1 D_1$ and $\alpha_3 > \theta_2 D_2$), (b) biological control of a weed ($\alpha_2 > \theta_1 D_1$ and $\alpha_3 < \theta_2 D_2$), (c) an alga–inefficient arthropod herbivore–efficient vertebrate predator ($\alpha_2 < \theta_1 D_1$ and $\alpha_3 > \theta_2 D_2$), and (d) a managed predator in a marine (e.g., phytoplankton–krill–whale) or terrestrial environment (e.g., plant–ungulate–predator) ($\alpha_2 < \theta_1 D_1$ and $\alpha_3 < \theta_2 D_2$). Note that only the direction of the inequalities and not the magnitude are implied. [Reprinted with permission from Gutierrez et al. 1994.]

Case 2: $(\alpha_2 > \theta_1 D_1$ and $\alpha_3 < \theta_2 D_2)$. In contrast, the dynamics depicted in Figure 8.4b represent the interaction of an inefficient predator and an efficient herbivore as might be expected of a successful weed biological control agent. The isocline inequalities are $\alpha_2 > \theta_1 D_1$ and $\alpha_3 < \theta_2 D_2$ suggesting a parabolic-shaped isocline for the weed and a sigmoidal isocline for the herbivore. For biological control to occur at low-plant density, the equilibrium point must be to the far left of the hump of the plant isocline, but it is unlikely to be stable (see Appendix II for stability arguments), thus creating a new paradox (Arditi and Berryman 1991). This paradox may, however, be answered by examining the case of the famous klamath weed where the chrysomelid beetle was so successful in the open range lands that the weed was pushed into less favorable shaded areas where it was not readily attacked (Huffaker and Kennett 1959). In these marginal areas, the resource base for the plant decreased, but the beetle's demand rate $(\theta_2 D_2)$ remained high resulting in lower environmental carrying capacities for both species. More important, the apparency of the weed to the beetle also declined enabling a stable interaction to occur because the stable point shifts to the right side of the plant isocline in marginal habitats. If the apparency of the weed to the beetle becomes low, a switch in the direction of the inequality may occur (i.e., $\alpha_2 < \theta_1 D_1$ and $\alpha_3 < \theta_2 D_2$) assuring stability in marginal areas (see Fig. 8.4d).

Case 3: $(\alpha_2 < \theta_1 D_1$ and $\alpha_3 \geq \theta_2 D_2)$. A fresh water aquatic systems consisting of an alga–arthropod herbivore–vertebrate predator (trout) might be illustrated by Figure 4c. From [13b] we see that the demand per unit biomass is greater for the alga than for higher trophic levels, and if the herbivores are inefficient grazers (low α_2) but the predator finds them efficiently (high α_3), the alga isocline must be asymptotic $(\alpha_2 < \theta_1 D_1)$ and that of the herbivore must be C-shaped $(\alpha_3 > \theta_2 D_2)$. If the alga–arthropod herbivore–fish system has isocline characteristics $\alpha_2 \ll \theta_1 D_1$ and $\alpha_3 > \theta_2 D_2$, overfishing might reduce predator density shifting the equilibrium causing an increase in herbivore biomass but having little effect on algal biomass. This might create a nuisance pest if the herbivore is a mosquito or blackfly.

Case 4: $(\alpha_2 < \theta_1 D_1$ and $\alpha_3 < \theta_2 D_2)$. In a marine system consisting of phytoplankton–krill–whale, the phytoplankton has a very high demand rate per unit biomass compared to krill, which has a high per unit demand compared to whale. Krill and whale are filter feeders and are assumed inefficient searchers (low α). This biology suggests that $\alpha_2 \ll \theta_1 D_1$ and $\alpha_3 \ll \theta_2 D_2$, and hence that the isoclines for phytoplankton and krill are asymptotic (Fig. 8.4d). This interaction predicts that the phytoplankton biomass is not affected much by the krill, and the krill is not much affected by the whale. However, exploitation of whales by humans (trophic level 4) makes them an intermediate trophic level with its isocline shifting to a C-shaped isocline as the relationship $\alpha_4 < \theta_3 D_3$ switches to $\alpha_4 \gg \theta_3 D_3$ as technology improves. The high α_4 value reflects our capacity to find and harvest whales, hence greatly increasing α_4 and D_4

might shift the equilibrium level of the whale to a small region around the saddle point increasing the possibility of extinction.

Moving down the trophic chain, the high productivity of krill makes their populations considerably more resilient to increasing exploitation (see Nicole and de la Mare 1993), but even here modern technology could adversely affect them and higher trophic levels as well. Factors, such as pollution, that adversely affect base resources lower the equiliburm densities of higher trophic levels in this and in other systems.

Nutritional Effects on Trophic Interactions

Increases in plant quality may lead to pest outbreaks if there is also an increase in the efficiency of conversion (θ_2) by the herbivore, causing a stable system with isocline characteristics $\alpha_2 > \theta_1 D_1$ and $\alpha_3 > \theta_2 D_2$ (Fig. 4a) to switch to an unstable one with isocline characteristics $\alpha_2 > \theta_1 D_1$, $\alpha_3 < \theta_2 D_2$ (Fig. 4b). In this case, an increasing θ_2 might cause the equilibrium point to shift sufficiently to the left of the hump of the plant isocline that instability results causing the pest to escape the control of its natural enemies or at least to cause the oscillations to greatly increase. Of course, as plant quality declines, θ_2 and the inequalities would switch again and stability would be restored. Examples of plant nutrient effects on pest outbreaks may include cyclical forest pests (e.g., spruce budworm, Peterman et al. 1979, Ludwig et al. 1978; larch bud moth, Fischlin and Baltensweiler 1979), increased outbreaks of mites on annual plants with increasing foliar nitrogen (e.g., Wermelinger et al. 1991b) and the influence of soil nutrients on the regional dynamics of pests (e.g., cassava, locusts, and armyworms; Gutierrez et al. 1988a; Neuenschwander et al. 1989; Janssen 1993).

FOOD WEBS

We can formulate and assess the dynamics of tritrophic food chains, but now we need to demonstrate that we can analyze an arbitrary food web where we have not only top-down and bottom-up effects but also lateral effects of competition, cannibalism, and other interactions (e.g., Fig. 8.5). DeAngelis et al. (1975) and Pimm (1982) used Lotka–Volterra models, and more recently Getz (1991), Berryman et al. (1995), and Arditi and Michalski (1995) used ratio dependent models to examine food webs. Arditi and Michalski state that in addition to homogeneity, models of food webs should be invariant. This means that if a species drops out of the food web, its effects are removed (case 1, Fig. 8.5), if a resource is the link between two consumer species, then its removal causes the food web to become two uncoupled systems (case 2, 8.5), and if two or more species are exactly identical then they reduce to one species (case 3, 8.5).

The ability to predict the consequences of such additions or deletions for

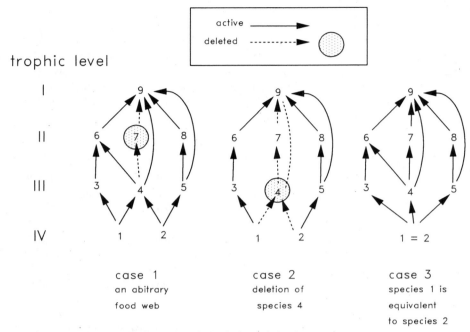

Figure 8.5. Examples of simple food web relationships (see text).

real systems remains an important challenge for resource managers. Simple intuitive criteria for deletion and persistence of species from food webs need to be developed if the role of a species in the food web is to be understood. A major difficulty with many food web models is that the underlying paradigm is often vague and the parameters have dubious interpretations making it difficult to relate their predictions to nature. An analysis of food web theory based on the metabolic pool model is given in Schreiber and Gutierrez (1996).

EPILOGUE

Despite the good qualitative predictions, we might be tempted to seek explanation beyond what is reasonable; especially for site specific evaluation. To answer these questions may require the development of a more detailed model. In general, incorporating the biology of age-specific resource acquisition and allocation in an age-structured population dynamics model does not represent a large conceptual leap; it is simply a bit more complicated. The number of parameters is roughly the same, but some are age dependent (see Chapters 9 and 10, and Gutierrez 1992).

9

Single Species Models with Age Structure

> *... x mature pigs = y tons of feeding-meal.*
> —**A. Milne, 1959**

In this chapter the population dynamics of single species models with age and mass structure are explored, and Nicholson's blowfly data are used to test the final model. Figure 9.1a depicts an age structured population having different stages (e.g., egg, larva, pupa and adult) and age classes ($i = 1, 2, \ldots, n$). All of the individuals are female and produce eggs that flow into the first age class as newborn, and all individuals in the ith age cohort corrected for survivorship flow into the $i + 1$th age class during the time interval t to $t + 1$. The basic number dynamics (N) may be described using a very simple difference equation model for births [1a]

$$N_o(t) = \sum_{i = i^*}^{n} rN_i(t) \qquad [1a]$$

and survival and aging [1b]

$$N_{i + 1}(t + 1) = N_i(t)l_{x_i}(t) \qquad [1b]$$

In this model, time (t) and age (i) are in the same units (days, physiological time, etc.), aging between cohorts is unity during the time step t to $t + 1$, $N_o(t)$ is the number of newborn produced by reproductive females of age i^* to maximum age n, r is the per capita reproductive rate, and lx_i is the proportion of individuals surviving in age group i and flowing to age group $i + 1$ during

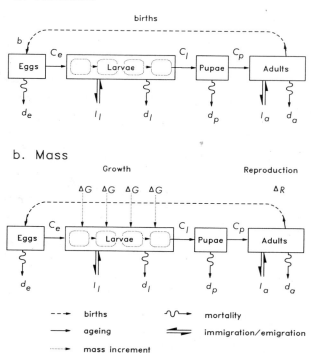

Figure 9.1. *A population model illustrating linked number and mass dynamics.*

time interval t to $t + 1$. Of course, the reproductive rate may be age-specific as well.

For convenience, we write a three age class matrix model version of this model (Leslie 1945) with the age-specific birth and survivorship rates in matrix **A**.

$$\begin{pmatrix} N_1 \\ N_2 \\ N_3 \end{pmatrix}_{t+1} = A \begin{pmatrix} N_1 \\ N_2 \\ N_3 \end{pmatrix}_t . \qquad [2a]$$

$$A = \begin{pmatrix} 0 & 0 & r \\ l_{x_1} & 0 & 0 \\ 0 & l_{x_2} & 0 \end{pmatrix} \qquad [2b]$$

Note that the model can be expanded to any number of age classes and sex ratios may be included with ease.

The continuous time equivalent of the Leslie model is the McKendrick (1926)–von Foerster (1959) model [3].

$$\frac{\partial N}{\partial t} + \frac{\partial N}{\partial a} = -\mu(t, a)N \quad \text{where} \quad \begin{cases} N(t, 0) = \int r(a)N(t, a)\, da \\ N(0, a) = v(a) \end{cases} \quad [3]$$

The net birth–death rate in [3] is μ, v is the initial population density, r is the age-specific per capita birth rate, and $N(t, a)$ is the population number-density function. Readers not familar with the relationship between models [2] and [3] should read Appendix II.

Oster and Ipaktchi (1978) modeled the number dynamics of Nicholson's blowflies using [3] incorporating density-dependent birth rates as well as functional dependence of fecundity on nutritional history. The model appeared to yield more of the subtleties of the dynamics than did prior models, but it did not provide significantly more understanding, due in part to the omission of the time-varying changes in mass dynamics (i.e., quality). Nicholson observed that the size of individuals in the population decreased as density related competition for food increased, and because nutritional history also affects per capita fecundity, longevity, and other life history traits, it is important to include the effects in the model (see Chapter 6).

ADDING MASS DYNAMICS

Adding time-varying mass dynamics to age-structure population models is not a difficult numerical problem, but it does complicate analytical investigations (Appendix II). A model for number (N) and mass (M) population dynamics for an insect such as blowfly is illustrated in Figure 9.1a and b. Numbers and mass are interchangeable units, hence we can formulate the dynamics in mass units and convert to numbers. The mass dynamics may be included in [4], where G_i and R_i are, respectively, the realized per capita age-specific growth and reproduction rates in mass units. As we shall see, G_i and R_i may be computed using the metabolic pool model.

Numbers (N)

$$N_1(t + 1) = N_0(t)$$

$$N_{i+1}(t + 1) = N_i(t)l_{x_i}(t) \quad [4a]$$

Mass (M)

$$M_1(t + 1) = R^*, \quad \text{where} \quad R^* = 0.5 \sum_{i=i^*}^{n} R_i N_i(t)$$

$$M_{i+1}(t + 1) = \{G_i(t)N_i(t) + M_i(t)\}lx_i(t) \quad \text{for} \quad 1 < i < i^* \quad [4b]$$

Only the details of the first age class are specified, and those of the other i classes are presented in a general form, indicating that some members of the population grow ($1 < i < i^*$) and one half of the adults are females of age $i^* \leq n$ who use the energy acquired for reproduction.

To estimate the average mass of individuals in each age cohort ($\overline{m}_i$), at any time t we simply divide M_i by N_i. The reproductive mass that flows into the first mass cohort is $R^*(t)$, and the number of newborn $N_0(t)$ equals

$$N_0(t) = \frac{R^*(t)}{\varepsilon \overline{m}} \quad [5a]$$

where the constant ε relates the mass of a newborn to the average mass of all mothers ($\overline{m}$).

$$\overline{m} = \frac{\sum_{i=1^*}^{n} M_i(t)}{\sum_{i=1^*}^{n} N_i(t)} \quad [5b]$$

The number and mass models [4a,b] including this biology may also be written as linked Leslie models (McKendrick–von Forester models) ([3]), but for simplicity, a Leslie model with three age classes (i.e., newborn, immatures, and adults) is presented with the number birth rate given as per capita values and mass birth and growth rates per unit mass.

$$\begin{pmatrix} N_1 \\ N_2 \\ N_3 \end{pmatrix}_{t+1} = \begin{pmatrix} 0 & 0 & \dfrac{R_3}{\varepsilon \overline{m}_3} \\ l_{x1} & 0 & 0 \\ 0 & l_{x2} & 0 \end{pmatrix} \begin{pmatrix} N_1 \\ N_2 \\ N_3 \end{pmatrix}_{t} \quad [6a]$$

$$\begin{pmatrix} M_1 \\ M_2 \\ M_3 \end{pmatrix}_{t+1} = \begin{pmatrix} 0 & 0 & R_3/\overline{m}_3 \\ l_{x1}(1 + G_1/\overline{m}_1) & 0 & 0 \\ 0 & l_{x2}(1 + G_2/\overline{m}_2) & 0 \end{pmatrix} \begin{pmatrix} M_1 \\ M_2 \\ M_3 \end{pmatrix}_{t} \quad [6b]$$

Note that the births and age-specific growth occur before mortality pointing to a potential artifact in model construction using difference equations (Wang and Gutierrez 1980). Links to other trophic levels in the dynamics models occur via age and time-varying parameters lx_i, R_3 and G_1 and G_2. Of course, the model can accommodate any number of age classes.

Model [7] is the continuous form of [6] incorporating attributes of time, age, and mass of the population (Sinko and Streifer 1967). If $da = dt$, then $N(t, a, m)$ satisfies

$$\frac{\partial N}{\partial t} + \frac{\partial N}{\partial a} + \frac{\partial[g(\cdot)N(t, a, m)]}{\partial m} = -\mu(\cdot)N(t, a, m) \qquad [7]$$

where $g(\cdot)$ and $\mu(\cdot)$ are the mass growth and death rate functions respectively. Equation [7] needs two initial conditions to complete its description of the population density function: $N(0, a, m)$ is the age–size distribution of the population at time 0, and $N(t, 0, m)$ is the size density function of newborn. We denote $N(0, a, m)$ and $N(t, 0, m)$ by $\alpha(a, m)$ and $\beta(t, m)$, respectively. Appendix II shows this equivalence and how we might discretize this model for programming on a computer.

VARIABLE MATURATION TIMES IN AGE-STRUCTURED MODELS

Not all individuals in a cohort develop at the same rate, and this may be introduced to a difference equation model [4] via an aging rate between age classes ($c < 1$, see Appendix II). For example, if we had only one age class (i.e., $n = 1$) with N individuals and a proportion $c = 0.1$ age out of the age class (dying in this case) each time step, then the resulting'distribution of exit times is a negative exponential with a time delay of zero. In a dynamics model with the same aging rate but more age classes, individuals of a cohort born at the same time enter the first stage with age zero and exit as deaths at maximum age. The distribution of emergence times (deaths) of members of a cohort is characterized by a delay and a mean and variance (see Appendix 9.1). Of course, the same arguments and model can be used for the completion of development in one stage and the transfer to the next life stage.

We can rewrite model [4] with age-specific mortality and aging rates between age classes included.

Numbers (N)

$$N_1(t+1) = (1 - c)N_1(t)lx_1(t) + N_o(t)$$

$$N_i(t+1) = (1 - c)N_i(t)lx_i(t) + cN_{i-1}lx_{i-1}(t) \qquad [7a]$$

Mass (M)

$$M_1(t + 1) = (1 - c)\{M_1(t) + G_1(t)N_1(t)\}lx_1(t) + R^*(t)$$

$$M_i(t + 1) = (1 - c)\{M_i(t) + G_i(t)N_i(t)\}lx_i(t)\}$$
$$+ c\{M_{i-1}(t) + G_{i-1}(t)N_{i-1}(t)\}lx_{i-1}(t) \qquad [7b]$$

The model assumes that survivorship occurs before aging and growth, but reproduction is unaffected within the time step. Whatever way we write this model, variants of these assumptions will occur (Wang and Gutierrez 1980). In this model, an infinitesimally small proportion of each cohort never emerge introducing an artifact into the dynamics. The *distributed delay* or *distributed maturation times* model (e.g., Goudriaan 1973; Manetsch 1976; Sharpe et al. 1977; Plant and Wilson 1986; Severini et al. 1990), properly formulated, avoids this problem. Gutierrez (1992) used the Manetsch (1976) model (and notation) to model the dynamics of Nicholson's laboratory populations of blowfly.

DISTRIBUTED MATURATION TIME MODELS

The distribution of maturation times in the absence of mortality is described by an Erlang distribution with parameter k equal to the number of age classes required to reproduce the observed mean and variance of the emergence times of an initial cohort. The theoretical basis for this model has been outlined by Severini et al. (1990, see Appendix 9.1). If we ignore mortality to the cohort, in theory k may be estimated from data as $k = \text{DEL}^2/\text{VAR}$ where DEL is the average developmental time and VAR is the variance.

The distributed delay model including age-specific time-varying mortality for the linked dynamics of k cohorts of numbers (N) and masses (M) may be written as follows:

Numbers (N)

$$\frac{dN_1}{dt} = x(t) - r_1(t) - \mu_1 N_1$$

$$\frac{dN_2}{dt} = r_1(t) - r_2(t) - \mu_2 N_2 \qquad [8a]$$

$$\vdots$$

$$\frac{dN_k}{dt} = r_{k-1}(t) - y(t) - \mu_k N_k$$

Mass (M)

$$\frac{dM_1}{dt} = x^*(t) - r_1^*(t) - \mu_1^* M_1$$

$$\frac{dM_2}{dt} = r_1^*(t) - r_2^*(t) - \mu_2^* M_2$$

$$\vdots$$

$$\frac{dM_k}{dt} = r_{k-1}^*(t) - y^*(t) - \mu_k^* M_k$$

where births of numbers (N) and mass (M) enter the population as $x(t)$ and $x^*(t)$, respectively, the net proportion age-specific mortality (i.e., death, immigration, emigration, and mass growth) during dt equals $-\infty < \mu_i < 1$ and $-\infty < \mu_i^* < 1$, individuals and their mass surviving to maximum age k exiting as $r_k(t) = y(t)$ and $r_k^*(t) = y(t)$ respectively. Note that μ and $\mu^* \to \infty$ if massive immigration of numbers and their associated mass occur. The number and mass in the ith age cohort equals $N_i(t) = r_i(t) \cdot \text{DEL}/k$ and $M_i(t) = r_i^*(t)\text{DEL}/k$, respectively, and the number and mass in the population equal $N(t) = \Sigma_{i=1}^{k} N_i(t)$ and $M(t) = \Sigma_{i=1}^{k} M_i(t)$, respectively. The instantaneous solution for the flow rates r_i and r_i^* are (Vansickle 1977):

Numbers (N)

$$\frac{dr_i}{dt} = \frac{k}{\text{DEL}} \left\{ r_{i-1}(t) - \left[1 + \mu_i(t) \frac{\text{DEL}}{k} \right] r_i(t) \right\} \qquad [9a]$$

Masses (M)

$$\frac{dr_i^*}{dt} = \frac{k}{\text{DEL}} \left\{ r_{i-1}'^*(t) - \left[1 + \mu_i^*(t) \frac{\text{DEL}}{k} \right] r_i^*(t) \right\}$$

The difficult trick in model development is to incorporate all biotic and abiotic factors affecting attrition in μ_i and μ_i^* defined as the proportional loss rates, and to write the computer algorithm to avoid computation errors. In the model, a 10% immigration rate to the ith cohort cancels a 10% death rate (i.e., $\mu = 0$), likewise a 10% mortality rate that reduces numbers may not result in the same reduction of cohort mass because the growth rate of the surviving individuals may be such that little or no reduction in total population biomass occurs. A PASCAL computer subroutine to accomplish this latter task is given in Appendix 9.2. This model is now applied to Nicholson's blowfly data.

A METABOLIC POOL-DISTRIBUTED DELAY MODEL OF NICHOLSON'S BLOWFLIES

In nature, biotic and abiotic factors affect birth, death and net immigration rates of species, and the intensity of the effects vary over time, however, in Nicholson's closed system experiment, all abiotic factors were controlled. In our model of blowfly, the age structure of number and mass (and hence the size structure) of the population and how the food supply affects the dynamics are incorporated in a distributed maturation time model. The metabolic pool model is used for the resource acquisition and allocation functions (cf. Gutierrez 1992).

The parameters of this model could not be determined precisely from Nicholson's data or derived from Readshaw and Cuff (1980) (see Chapter 6) because all values were given per fly. As Nicholson clearly stated—the size of flies (i.e., and hence their demand) varied over the time of the experiment, hence, to capture these dynamics in our model, the parameters must be put on a milligram per milligram basis. The following guesses proved reasonable: maximum fly size is 50 mg biomass (m_{max}); egg size is 0.1 mg; maximum fecundity is 10 eggs female^{-1}day^{-1} (Readshaw and Cuff 1980); the respiration rate [$z(\cdot)$] is 0.3 liver mg^{-1} day^{-1}, the proportion egested is ($1 - \beta = 0.1$); 2, 6, 5, 2, and 6 days are the observed egg, larval, pupal, preoviposition, and mature adult periods, respectively. A wide range of values for Erlang parameter $k \in [30, 50]$ gave similar results, hence 40 was used.

Some important biological observations are relevant here. In many insect species, total fecundity is known to be related to female size ($m(t, a)$) (e.g., bush fly, Tyndale-Briscoe and Hughes 1969), and hence in our model, fecundity was corrected by the ratio of current size minus the minimum size of viable adult (10 mg) to the maximum size.

$$0 < \frac{m(t, a) - 10 \text{ mg}}{m_{max}} < 1 \qquad [10]$$

Nicholson made the critical observation that very small mature larvae failed to pupate, accounting for the size threshold of 10 mg in the model. In this closed system, emigration did not occur, and larvae were forced to compete for the available food (i.e., *scramble competition*). In laboratory studies, Baumgärtner et al. (1981) observed, in the absence of competition, reductions in size for three species of predators across levels of food, but survivorship was unaffected except at very low food levels (see Readshaw and Cuff 1980). In an open system, individuals receiving insufficient food would emigrate at a rate proportional to the supply/demand ratio.

SIMULATION STUDIES

Only Nicholson's experiments utilizing constant larval food (25 or 50 g day^{-1}) and ad lib adult food were used in this study. No attempt was made to model experiments where adult food was constant and larval food was ad lib because this required additional assumptions. For consistency of units, the liver provided as food was converted to dry biomass assuming a water content of 70%. The model was started using 100 adults and was run for 500 days. As in Nicholson's reports, only the data for days 240–440 are shown (Figs. 9.2 and 9.3). These are reasonable estimates and assumptions as anyone reading Nicholson's experimental methods would be hard pressed to reconstruct the regimes and the initial conditions accurately.

The simulations produced reasonable fits to the data, but they were not significantly better than those reported by prior workers. Important differences to prior studies are that the larval supply/demand ratio (Figs. 9.2b–9.3b) ex-

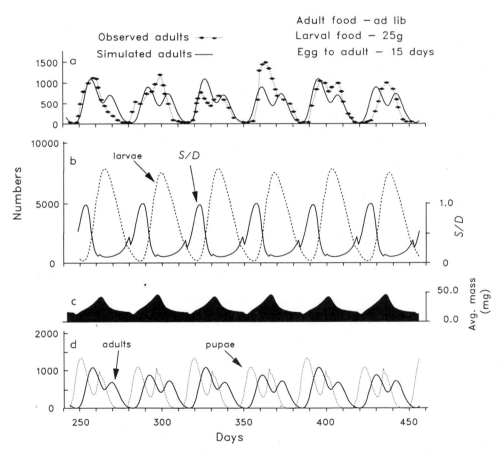

Figure 9.2. Simulation of Nicholson's blowfly data for the 25-g larval food and ad lib adult food treatment using the metabolic pool approach for the acquisition and allocation functions and the distributed maturation time model to simulate the population dynamics: (a) observed and simulated data, (b) simulated larval population and the resulting supply demand ratio, (c) average adult mass, and (d) simulated adult and pupal numbers to illustrate the origins of the adult peaks [see Gutierrez 1992].

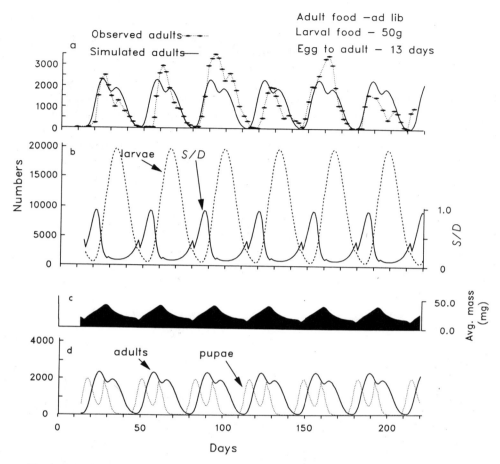

Figure 9.3. *Simulation of Nicholson's blowfly data for the 50-g larval food and ad lib adult food treatment using the metabolic pool approach for the acquisition and allocation functions and the distributed maturation time model to simulate the population dynamics: (a) observed and simulated data, (b) simulated larval population and the resulting supply demand ratio, (c) average adult mass, and (d) simulated adult and pupal numbers to illustrate the origins of the adult peaks [see, Gutierrez 1992].*

plains most of the underlying adult dynamics and the magnitudes and sequence of the observed and simulated peaks are not uniform as predicted by prior models. The ratio decreases when larval numbers increase indicating competition for food, low per capita growth rates, and poor pupation success of small larvae. The effects of the latter mortality is seen as the double peaks in subfigures 9.2d and 9.3d.

How can we further explain these discrepancies without having to invoke novel theories? Anyone who has conducted experiments with insects knows procedures vary and measurements are not always made at exactly the same time. I would also suspect that in Nicholson's experiments the liver was not always given at the same time each day, its water content and nutritional value were not constant, and no doubt other experimental mishaps occurred (e.g., the lads may have had a few pints on Christmas, New Years, and or course St. Patrick's day). Since food was often limiting for the larvae, this would be akin to providing variable rates of food. We simulated this variability by introducing a stochasticity scalar to the daily food supply using a uniform probability function with an expected mean of 1.0 and a range of 0.5–1.5.

The results of the stochastic simulation of the 50-g larval food experiment

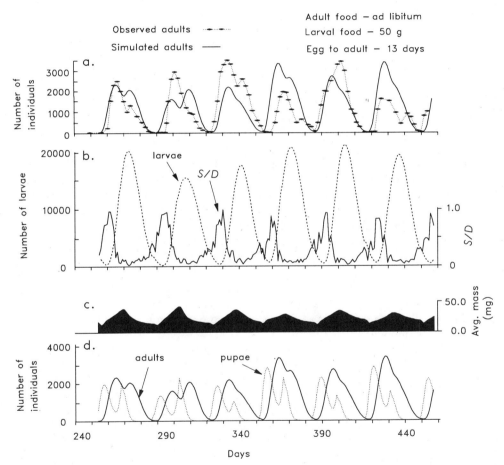

Figure 9.4. *Stochastic simulation of Nicholson's blowfly data for the 50-g larval food and ad lib adult food treatment using the metabolic pool approach for the acquisition and allocation functions and the distributed maturation time model to simulate the population dynamics: (a) observed and simulated data, (b) simulated larval population and the resulting supply demand ratio, (c) average adult mass, and (d) simulated adult and pupal numbers to illustrate the origins of the adult peaks. [Reprinted with permission from Gutierrez 1992.]*

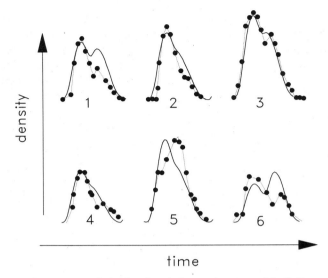

Figure 9.5. *Comparison of simulated and observed patterns [cf., Gutierrez 1992].*

are shown in Figure 9.4 (Gutierrez 1992). The population cycles remain unchanged as to period, but the magnitudes of the peaks are highly variable. In general, the model mimics the variability of the data but not the sequence of the peaks. This is to be expected as we do not know the pattern or magnitude of the variability and of course errors propagate over time. We note similarity of observed (O) and simulated (S) peaks: O(1) to S(3), O(2) to S(5), O(3) to S(4), O(5) to S(6), and O(6) to S(1) (see Fig. 9.5). What this suggests is that variability in the pattern and quantity of the food supplied is sufficient to explain, but not to predict, the observed data because we do not know the pattern of the variability. Furthermore, despite the stochasticity of the food supply, all of the underlying patterns except the short run supply/demand ratio remained smooth.

Thus, Nicholson's experiments are merely a variant of something taken for granted in animal husbandry: that one can predict the rate of conversion of the rations given to livestock to yields. But this same point was made by Milne (1959) in a reply concerning Nicholson's model: . . . *x mature pigs = y tons of feeding-meal.* The population dynamics we observe in nature are but variants of this problem: weather and other abiotic factors determine the maximal expression of potential population growth, and biotic interactions operating on this background yield the resulting dynamics.

EPILOGUE

The origins of single species models were outlined in Chapter 7 and their predictions have been compared using Nicholson's blowfly data as a common foil there and in this chapter. Here we added the realism of population age and mass structure, of distributed maturation times and the metabolic pool model. This was not difficult, and the model proved sufficient to explain the observed patterns in the dynamics data. Adding age-structure to models is not new to ecology, and certainly using the metabolic pool approach to reproduce the data starting from first principles is not new to us. The ability to simulate the blowfly dynamics is not fortuitous and demonstrates the utility this meta-physiological approach. Approaching the population dynamics problem in this manner makes the model independent of the data, and is the major progress reported here. In Chapter 10, we examine multitrophic applications of real world problems building on the model developed here.

Some Principles of Population Development and the Time Distributed Delay Models (cf., Severini et al. 1990)

The ability to forecast phenological events has been an important agricultural activity for a long time. In supervised pest control, for example, the decision making depends on the prediction of key events such as the occurrence of adult pest insects and their subsequent reproduction. Welch et al. (1978) found that Manetsch's (1976) time distributed delay model was highly appropriate for phenological purposes and used it as the core for computer-based decision making. However, to account in a concise way for losses and gains during population development, Gutierrez and Baumgärtner (1984b) incorporated Vansickle's (1977) attrition into the model (i.e., population dynamics). This permitted the transition from phenology models to population models and allowed the structuring of complicated multitrophic systems models by adding the metabolic pool model (e.g., Gutierrez et al. 1984b, 1987, 1988b).

Recently, however, it became clear that the time distributed delay model was more than a convenient way to describe the cohort development in demographic analyses. Severini et al. (1990) indicate that the model represents the process underlying the development of cohorts. In this chapter the most important elements are summarized, but readers are referred to the original publications for complete details.

THEORETICAL CONSIDERATION OF DISTRIBUTED DELAY MODELS WITHOUT ATTRITION

All individuals of a population pass through different phenophases or life stages to complete the immature development (arthropods) or to reach the end of their life span (plant subunits or arthropods). Suppose there are $k = 1, 2, \ldots, K$

stages in a life cycle, and that the passage of a population through them can be represented by a boxchain.

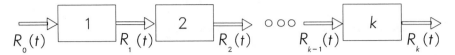

Figure A9.1. *The flow dynamics between stages. [Redrawn from Severini et al. 1990.]*

The $R_k(t)$ is the output flux from the kth stage, and $R_o(t)$ and $R_K(t)$ are input and output fluxes of the whole cycle, respectively. The number K of the cycle stages depends on the population biology, but the generic kth box has always the same logical structure: an input $R_{k-1}(t)$ and an output $R_k(t)$, as shown.

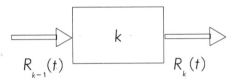

Figure A9.2. *The flow dynamics in the last stage cohort. [Redrawn from Severini et al. 1990.]*

They assumed that a linear differential operator **L** of order **H** determines the relationship between input and output of a generic box:

$$\mathbf{L}[R_k(t)] = R_{k-1}(t) \tag{A9.1}$$

with

$$\mathbf{L} = \alpha_0(t)\frac{d^{(H)}}{dt^{(H)}} + \alpha_1(t)\frac{d^{(H-1)}}{dt^{(H-1)}} + \cdots + \alpha_{H-1}(t)\frac{d}{dt} + \alpha_H(t) \tag{A9.2}$$

The operator coefficients α_h with $h = 0, 1, 2, \ldots, H$ in poikilotherm populations must depend on time t through temperature T:

$$\alpha_h(t) = \alpha_h[T(t)] \tag{A9.3}$$

In the particular situations of development at constant temperature, **L** becomes a linear operator with constant coefficients:

$$L_T = \alpha_o\frac{d^{(H)}}{dt^H} + \alpha_1\frac{d^{(H-1)}}{dt^{H-1}} + \cdots + \alpha_{H-1}\frac{d}{dt} + \alpha_H \tag{A9.4}$$

The linear equation of order **H** arising from the application of the operator **L_T**, that is,

$$L_T[R_k(t)] = R_{k-1}(t) \qquad \text{[A9.5]}$$

is equivalent to a system of **H** linear equations of first order

$$\frac{dr_h(t)}{dt} = C_h[r_h(t) - r_{h-1}(t)] \qquad \text{[A9.6]}$$

$$h = 1, 2, \ldots, H$$

where $r_o(t) = R_{k-1}(t)$ and $r_H(t) = R_k(t)$, and the remaining $r_h(t)$ are called intermediate fluxes; according to the graphical representation.

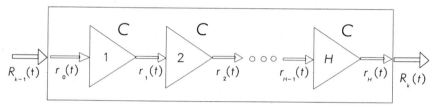

Figure A9.3. Flow dynamics between substages (see Fig. A9.2). [Redrawn from Severini et al. 1990].

In the so-called continuous delay approach, all the H substages are considered as identical,

$$C_h = C \qquad \text{[A9.7]}$$

and are represented by the following time invariant delay system of equations:

$$\frac{dr_h(t)}{dt} = C[r_h(t) - r_{h-1}(t)] \qquad \text{[A9.8]}$$

where the number of equations H must be determined together with the value of C.

System [A9.8] cna be solved analytically. For example, if the input flux $r_o(t)$ is a Dirac Delta function

$$R_{k-1}(t) = \delta(t - t_0) = \begin{cases} = 0 & \text{for} \quad t \neq t_0 \\ \neq 0 & \text{for} \quad t = t_0 \\ \text{and such that} \quad \int_{+\infty}^{-\infty} \delta(t)\, dt = 1 \end{cases} \qquad \text{[A9.9]}$$

and this time invariant delay system can easily be solved by the method of Laplace Transform, giving as a result:

$$r_h(t) = \frac{C_H}{(H-1)!} e^{-C_t H - 1_t} \qquad [\text{A9.10}]$$

which is generally known as the Hth function of the Erlang's Family Distribution. It contains a parameter H for the shape and C for the scale of the function.

Since each function is normalized, it can be considered as the probability distribution of the transit times. By definition, the expected value DEL of this distribution, that is, the expected delay between input and output, and its variance VAR are

$$\text{DEL} = \frac{C_H}{(H-1)!} \int_0^\infty e^{-C_t} t^H \, dt = \frac{H}{C} \qquad [\text{A9.11}]$$

$$\text{VAR} = \frac{C^H}{(H-1)!} \int_0^\infty (\text{DEL} - t)^2 e^{-C \cdot t} t^{H-1} \, dt = \frac{H}{C^2} \qquad [\text{A9.12}]$$

From a biological point of view, the Dirac Delta function can be regarded as a cohort whose statistics DEL and VAR can be studied under laboratory conditions (Gilbert et al. 1974; Welch et al. 1978) and, with some constraints, under field conditions (Severini et al. 1990). From [A9.11] and [A9.12] these experiments yield

$$C = \frac{H}{\text{DEL}} \qquad [\text{A9.13}]$$

which can be substituted into [A9.7] to yield

$$\frac{dr_h(t)}{dt} = \frac{H}{\text{DEL}} [r_h(t) - r_{h-1}(t)] \; h = 1, 2, \cdots, H \qquad [\text{A9.14}]$$

which is usually presented as the time invariant delay system (Manetsch 1976).

PRACTICAL CONSIDERATIONS

Severini et al. (1990) concluded that the cohort is the appropriate population unit that permits the analysis of demographic processes. As a consequence, population development can be studied via a series of cohorts entering one or several consecutive life stages or phenophases (Manetsch 1976; Welch et al. 1978; Gutierrez and Baumgärtner 1984b). Severini et al. (1990) discuss different parameter estimation procedures.

The theory as presented here assumes a constant temperature regime, and modifications are required to account for poikilothermic development under fluctuating temperatures. This result can be achieved in two different ways. In the first and more widely used approach DEL_T can be considered as a thermal constant, that is, the transit time through a life stage or phenophase can be expressed as a constant having physiological units such as day-degrees (Manetsch 1976; Welch et al. 1978; Gutierrez et al. 1984). This method is the time-invariant approach. As a result, the population and its cohorts travel on a physiological time horizon that substitutes for calendar time (Gilbert et al. 1976; Gutierrez and Baumgärtner 1984b; Gutierrez et al. 1984b). If necessary, this constant can be modified to account for nutritional effects (Gutierrez et al. 1984b, 1988b).

In the second approach DEL_T becomes time-varying ($DEL\{T(t)\}$) to account for the effects of temperature and other environmental variables on transit times on a calender time scale. Time-varying distributed delays have been applied by Baumgärtner and Severini (1987) to insects and by Tamò and Baumgärtner (1993) to plants (see also Curry and Feldman 1987).

MODELS WITH ATTRITION

The above theory has been developed for passage through life stages or phenophases in which no losses occur. The resulting model is therefore only applicable to demographic processes describing phenology (Welch et al. 1978; Baumgärtner and Severini, 1987). As soon as mortalities, migration patterns, or growth are taken into consideration [A9.14] is no longer satisfactory. Vansickle (1977) developed a method for incorporating attrition into both time invariant and time-varying delay models. His method permitted, as briefly described by Gutierrez et al. (1988b), Baumgärtner and Gutierrez (1989) and by Baumgärtner and Bonato (1991), the transition from the above phenology models to single-species population models (e.g., Zahner and Baumgärtner 1988; Baumgärtner et al. 1990; Bianchi et al. 1990; Cerutti et al. 1991), to crop canopy models (Gutierrez et al. 1984b, 1987, 1988b; Wermelinger et al. 1991a; Tamò and Baumgärtner, 1993), to multitrophic population models (Gutierrez et al. 1984a, 1987, 1988a) and last to an object oriented multitrophic population model (Gutierrez and Ellis unpublished).

Pascal Subroutine for Distributed Maturation Times With and Without Attrition

```
Procedure Delay (Vin:single; var Vout,Shed:single; var R:single50;
             Plr: single50; del:single; K:integer; Dt:single);
{Delay subroutine to age population and account for attrition }
(*
    Vin{input increment}
    Vout          {flow out of R array}
    Shed          {attrition from array}
    R             {R array}
    Plr           {Attrition array, i.e. proportional net mortality rate }
    Del           {Mean time through the delay process, i.e., longevity}
    K             {number of substages in R}
    Dt            {Amount of time to process, units of physiological time,
                  days,etc}
*)
var
    i, j, idt : integer;
    a, shd,x : single;
begin
    Shed := 0.0;
    Vout := 0.0;
    if Dt>0.0 then
    begin
        Idt := trunc(1.0 + (2.0*Dt*(K/Del)));
        A := (Dt / (Del/K)) / Idt;
      { A = proportion flow rate from one substage
                          to next }
```

157

```
if(Nonzero(Plr,K))then
{Plr is nonzero = attrition is done here}
begin
    x:=del/(k*dt);
    { x corrects plr for dt compared to Del/K}
    for i:= 1 to k do plr[i]:=1.0-power(1.0-plr[i],x);
    for J := 1 to ldt do
    begin
        Vout := Vout + A*R[K];
        for i := K downto 2 do
        begin
            R[i]= R[i] + A*(R[i-1]-R[i]);
            Shd := (1.0-power(1.0-plr[i],A))*R[i];
            Shed := Shed+Shd;
            R[i] := R[i]-Shd;
        end;
        if(J = 1) then R[1] := R[1]+A*((Vin/Dt)*ldt-R[1])
        else R[1] := R[1] - A*R[1];
        Shd := (1.0-power(1.0-plr[i],A))*R[1];
        Shed := Shed + Shd;
        R[1] := R[1] - Shd;
    end {For J, with attrition}
else


    {Plr is zero = no attrition is done }
    for j := 1 to idt do
    begin
        Vout : Vout + A*R[K];
        for i := K downto 2 do R[i] := R[i] + A*(R[i-1] - R[i]);
        R[1] := R[1] + A*((Vin/Dt) - R[1]);
    end; {for j}
end; {If Dt > 0.0}
end; { Procedure Delay}
```

Functions used in the delay subroutine

```
Function Power(base,exponent : real): real;
(* exponentiation : power := base**exponent *)
begin
 if base=0.0 then power:=0.0 else
 Power:=exp(exponent * ln(base))
end; {Power}
```

```
Function Sum(Var a : single50; m,n : Integer ) : single;
(*
Sum the entries of an array.
m and n must be < = constant K.
*)
Var I : integer;
    Temp : single;
Begin
 temp:=0.0;
 For i := m to n do temp:= temp + a[i];
 Sum:=temp;
end; {function Sum}



Function Nonzero(a:single50; k:integer) : Boolean;
{ Nonzero is true if any element of array A is nonzero.}
var
    i:integer;
begin
    i:=1;
    while (i<k+1) and (a[i]=0.0) do Inc(i);
    Nonzero := (i<k+1);
end;
```

10

Realistic Age-Structured Multitrophic Models

A model cannot capture every detail we observe in nature, because if it did it would be as difficult to understand as nature itself. Unfortunately, the devil is often in the details.
—Anonymous

The addition of age structure and other aspects of biological realism to single species models was addressed in Chapter 9, and is extended here to trophic interactions. As before, analogous processes are assumed at all trophic levels. Extending the model to include more complicated biology, or using a different dynamics model is not conceptually difficult once the basic model is understood. In the first section, we add realism in a stepwise fashion. We use difference equation models for this purpose because it is easier to show how the biology is included. In the next section we use the continuous analogue of these models to show how the dynamics of modular populations (e.g., fruits, embryos, etc.) are included in an overall population dynamics model. The examples, including the object oriented examples, given in the last section are based on the distributed maturation time model that has been widely used for modeling field systems (see Chapter 9, and review Appendix II).

A SIMPLE BITROPHIC MODEL

In a simple age-structured two-species difference equation model, [1, 2], of the number dynamics of a predator–prey system (i.e., in the broadest sense), stage 1 of the prey (N_1), is attacked by stage 2 of the predator (P_2).

Prey   ·[1]

$$N_1(t + 1) = rN_2(t) \qquad \text{(Immatures)}$$

$$N_2(t + 1) = f\left(\frac{P_2(t)}{N_1(t)}\right)N_1(t) \qquad \text{(Adults)}$$

Predator   [2]

$$P_1(t + 1) = h\left\{\left(1 - f\left[\frac{P_2(t)}{N_1(t)}\right]\right)\right\}N_1(t) \qquad \text{(Immatures)}$$

$$P_2(t + 1) = P_1(t) \qquad \text{(Adults)}$$

Models [1] and [2] assume that all mortality is due to predation and that [1 − $f(P_2(t)/N_1(t)$] is a type II ratio dependent random search functional response model that estimates the proportion of N_1 captured by P_2, r is the constant per capita prey birth rate, and h is the conversion rate of prey captured to predator progeny. For the prey, all individuals in age class 2 die and survivors of predation in age class 1 transfer to age class 2 during the time period t to $t +$ 1. Similarly, for the predator, all individuals in age class 2 die, those in age class 1 transfer to age class 2, and prey attacked modified by h enter the first age class. If the time and age steps are equal and in the same units, equations [1] and [2] can be modeled as linked Leslie (1945) models [3].

Prey

$$\begin{pmatrix} N_1 \\ N_2 \end{pmatrix}_{t+1} = \begin{pmatrix} 0 & r \\ f(\cdot) & 0 \end{pmatrix}\begin{pmatrix} N_1 \\ N_2 \end{pmatrix}_t \qquad [3a]$$

Predator

$$\begin{pmatrix} P_1 \\ P_2 \end{pmatrix}_{t+1} = \begin{pmatrix} 0 & h[(1 - f(\cdot)]N_1/P_2 \\ 1 & 0 \end{pmatrix}\begin{pmatrix} P_1 \\ P_2 \end{pmatrix}_t \qquad [3b]$$

Note that in [1, 2, 3a,b], the prey and predator birth rates are per capita and prey aging occurs after predation. The ordering of biological processes is an important consideration in the development of difference equation models (Wang and Gutierrez 1980).

ORDERING OF BIOLOGICAL PROCESSES

If we introduce the simplest notion of variable aging via coefficients $c_i \in [0,$ 1], $i = 1, 2, 3, 4$, only part of each cohort ages to the next age class during the time step ([4] and [5], Figure 10.1) forcing further adjustments in ordering

a. Survivorship before aging

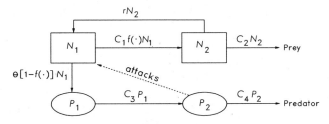

b. Aging before survivorship

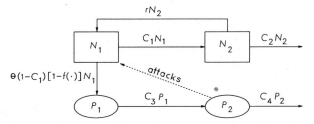

Figure 10.1. *Age structure predator (P) and prey (N) interaction: (a) survivorship before aging, and (b) aging before survivorship. The dashed line indicates the flow of information and the solid line the flow of numbers between and within species.*

the sequence of predation and aging. To simplify the notation, we define $f(\cdot) = f[P_2(t)/N_1(t)]$.

Predation before aging (Fig. 10.1a). [4]

$$N_1(t + 1) = (1 - c_1)f(\cdot)N_1(t) + rN_2(t) \qquad \text{(Immatures)}$$

$$N_2(t + 1) = (1 - c_2)N_2(t) + c_1f(\cdot)N_1(t) \qquad \text{(Adults)}$$

$$P_1(t + 1) = (1 - c_3)P_1(t) + h\{1 - f(\cdot)\}N_1(t) \qquad \text{(Immatures)}$$

$$P_2(t + 1) = (1 - c_4)P_2(t) + c_3P_1(t) \qquad \text{(Adults)}$$

Aging before predation (Figure 10.1b). [5]

$$N_1(t + 1) = (1 - c_1)f(\cdot)N_1(t) + rN_2(t) \qquad \text{(Immatures)}$$

$$N_2(t + 1) = (1 - c_2)N_2(t) + c_1N_1(t) \qquad \text{(Adults)}$$

$$P_1(t + 1) = (1 - c_3)P_1(t) + h(1 - c_1)\{1 - f(\cdot)\}N_1(t) \qquad \text{(Immatures)}$$

$$P_2(t + 1) = (1 - c_4)P_2(t) + c_3P_1(t) \qquad \text{(Adults)}$$

A prey refuge is created in [5] (Figure 10.1b) but not in [4] (Figure 10.1a) as some prey age out of the susceptible stage before predation occurs (i.e., aging before predation). A continuous model would solve this problem as everything occurs instantaneously, but this setting might prove inappropriate for modeling discrete phenomena caused by seasonality, behavior, or other factors. Note that if too much complexity is added, a differential model may not have a closed form solution, and hence would have to be discretized and evaluated numerically.

ADDING MASS STRUCTURE

Populations also have mass age structure, since the young are commonly smaller than the old and the average size within an age class may vary over time. True predators attack individual prey, but their consumption rate is related to the mass of the prey and gut size. In contrast, parasitoids also attack individuals, but the mass of the prey is not a limiting factor unless it is of the wrong size and is rejected for behavioral reasons (i.e., preference). For parasitoids, the egg load determines the demand for hosts.

Let us consider the biology of a *koinobiont* parasitoid whose hosts continue to feed after parasitism. To show how age-structured mass dynamics might be included in this model, consider the extension of [4] to include host mass (M) and parasitoid mass (Q) for the case of parasitism before aging in the host, [6], and aging before reproduction in the parasitoid, [7] (see Fig. 10.2). In this case we would use the parasitoid form of $f(\cdot)$, say the Fraser-Gilbert model, (F-G model, see Appendix 5.1).

Host [6]

$$N_1(t + 1) = (1 - c_1)f(\cdot)N_1(t) + rN_2(t)$$

$$N_2(t + 1) = (1 - c_2)N_2(t) + c_1f(\cdot)N_1(t)$$

$$M_1(t + 1) = (1 - c_1)f(\cdot)\{M_1(t) + G_1N_1(t)\} + m_N rN_2(t)$$

$$M_2(t + 1) = (1 - c_2)M_2(t) + c_1f(\cdot)M_1(t)$$

Parasitoid [7]

$$P_1(t + 1) = (1 - c_3)P_1(t) + h\{1 - f(\cdot)\}N_1(t)$$

$$P_2(t + 1) = (1 - c_4)P_2(t) + c_3P_1(t)$$

$$Q_1(t + 1) = (1 - c_3)(Q_1(t) + G_1P_1(t)) + m_p h\{1 - f(\cdot)\}N_1(t)$$

$$Q_2(t + 1) = (1 - c_4)Q_2(t) + c_3Q_1(t)$$

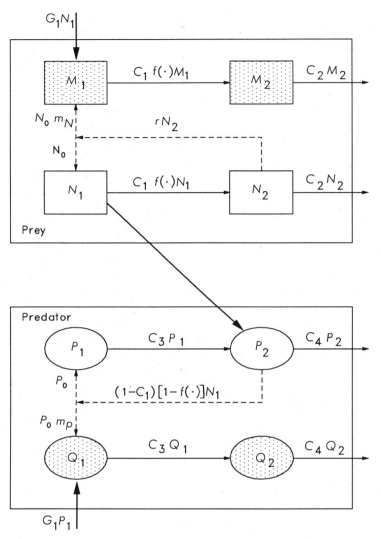

Figure 10.2. An age-structured two species predator–prey interaction model illustrating the dynamics of both numbers and biomass (shaded).

The function $f(\cdot)$ is in number units and is the proportion of hosts escaping attacked, while $(1 - f(\cdot))$ is the functional response model linking host death rate and parasitoid birth rate. Note that mortality within a stage affects numbers and mass equally. The birth rates of N and P are $N_0(t) = rN_2(t)$ and $P_0(t) = [(1 - f(\cdot)]N_1(t)$ in number units respectively, and $m_N = m_p$ converts them to mass units ($m_N N_0$ and $m_p P_0$). Hosts attacked by parasitoids (P_0) are transferred to stage P_1 in the parasitoid number dynamics model, and their mass ($m_p P_0$)

to Q_1 in the mass dynamics model. The mass of life stages N_1 and P_1 grow at a constant per capita rate G_1, but in nature rates would vary over time (see below). This model merits close attention before proceeding to other sections. To check your understanding, how would you change the model if P was a true predator? Hint: $f(\cdot)$ would be expressed in mass units because $f(P_2(t)/N_1(t)) \neq f(M_1(t)/Q_2(t))$.

ADDING THE METABOLIC POOL MODEL AND THE EFFECTS OF SUPPLY–DEMAND MORTALITY

Adding the biology of resource acquisition and allocation to growth and reproduction and the effects on survivorship is not difficult (see also Chapter 6). In the model below, $1 - f(\cdot)$ still describes the biology of parasitoid attack, and what we need to define are the dynamics of mass growth by immature stages and reproduction by host adults $(1 - g(\cdot))$, and the effects of supply/demand on survivorship.

Assume in [6, 7] that parasitized (P_1) and healthy (N_1) hosts feed on the same constant resource (R), and for simplicity their time varying realized growth rates both equal $(G_1(t))$. If N_2 also feeds on R, then we define its mass birth rate as $G_2(t)$. One way to include this time varying biology is to substitute the metabolic pool model (G-B model, see Appendix 5.1) in [6] and [7] for G_1 and G_2 which are redefined as follows.

$$G_1(t) = \theta \left[\left(1 - g \left(\frac{R}{(D_1(N_1 + P_1) + D_2 N_2)} \right) \right) \right] D_1 - z_{N_1}$$

$$G_2(t) = \theta \left[1 - g \left(\frac{R}{(D_1(N_1 + P_1) + D_2 N_2)} \right) \right] D_2 - z_{N_2}$$

[8]

The per capita mass demand rate for both parasitized and healthy hosts (N_1 and P_1) is D_1 (see Chapter 6) and the demand rate of adult hosts is $D_2 = m_N r$ (see [6]). Because these life stages are competing for the same resource, their demands are included in the denominator of $g(\cdot)$ as $D(t) = D_1(N_1(t) + P_1(t) + D_2 N_2(t)$, and used to estimate the realized acquisition rates (see Chapter 6). Respiration (i.e., $z_{N_i} = z_{P_1}, z_{N_2}$) and conversion costs ($\theta$) costs are subtracted from the resource acquired to yield time varying rates of per capita growth $[G_1(t)]$ and reproduction $[G_2(t)]$.

If the per capita supply $S(t)$ is $[1 - g(R/D)] D < D$, shortfalls of essential resources at time t affect emigration and death rates, but in our model we do not distinguish between them. A survivorship function based on the supply/demand ratio $0 \leq \phi(t) = [S(t)/D(t)] = 1 - g(\cdot) < 1$ is used to estimate the net loss.

We introduce $G_1(t)$, $G_2(t)$, and $\phi(t)$ in our model as follows:

Host [9]

$$N_1(t + 1) = \{(1 - c_1)f(\cdot)[N_1(t) + G_2(t)N_2(t)/m_N]\} \cdot \phi(t)$$

$$N_2(t + 1) = \{(1 - c_2)N_2(t) + c_1f(\cdot)N_1(t)\} \cdot \phi(t)$$

$$M_1(t + 1) = \{(1 - c_1)f(\cdot)[M_1(t) + G_1(t)N_1(t) + G_2(t)N_2(t)]\} \cdot \phi(t)$$

$$M_2(t + 1) = \{(1 - c)M_2(t) + c\,f(\cdot)[M_1(t) + G_1(t)N_1(t)]\} \cdot \phi(t)$$

Parasitoid [10]

$$P_1(t + 1) = \{(1 - c_3)P_2(t) + h(1 - f(\cdot))N_1(t)\} \cdot \phi(t)$$

$$P_2(t + 1) = (1 - c_4)P_2(t) + c_3P_1(t) \cdot \phi(t)$$

$$Q_1(t + 1) = \{(1 - c_3)[Q_1(t) + G_1(t)P_1(t)] + h(1 - f(\cdot))M_1(t)\} \cdot \phi(t)$$

$$Q_2(t + 1) = (1 - c_4)Q_2(t) + c_3[Q_1(t) + G_1(t)P_1(t)] \cdot \phi(t)$$

In summary, we have assumed the biology of a koinobiont parasitoid, both species produce only females, both species have only two age classes, interactions of parasitized and healthy hosts with the resource are in mass units $(g(\cdot))$ and those of the parasitoid with the host are in number units $(f(\cdot))$. The ordering of biological events in the prey model is parasitism, reproduction and growth, aging and S/D survivorship. In the parasitoid model, the order is reproduction and growth, aging and S/D survivorship on stages that feed on R. All models embed such assumptions and they require careful examination.

We could have formulated our model in several ways, and each would affect the model's realism, predictions and mathematical properties. In formulating a model, we must identify age-dependent trophic relationships and be clear about the assumptions we are making. Other biologies and other attributes such as more stages and ages, sex, morph, nutrient status, and so on, are not conceptually difficult to add given that [9] and [10] are fully understood. Building complicated models is relatively easy, what is difficult is the analysis of a model's properties (see Appendix III). As an exercise, the reader is asked to consider how the model would change: if the parasitoid were an *idiobiont* parasitoid and its host stopped feeding after parasitism; if we were dealing with a true predator; if the sex ratios were less than 1). In the next section, this model is extended to modular organisms.

MODELING MODULAR POPULATIONS (POPULATIONS WITHIN POPULATIONS)

Individual plants in an age-structured population have within them age-structured subunit populations of leaves, stem, root, and fruits (Harper and White 1974; Gutierrez et al. 1975; Gutierrez and Wang 1976; Law 1983). The subunit populations within individual plants are linked via the dynamics of resource supply–demand (i.e., the elements of the metabolic pool). Similarly, female animals may have age-structured populations of ova and/or embryos developing within them that are also linked via resource dynamics. Figures 10.3 and 10.4 present conceptual models of populations of plants and of animals having subunit populations. Intrinsic birth and death rates of subunits within individuals or whole individuals at the population level are modified by the factors that reduce the supply/demand ratio. Of course abiotic factors (e.g., heat or cold) and biotic factors (e.g., predation) may affect the numbers (and masses) of organisms directly, and death of parents (usually) kills the subunits they contain. The models required to describe the dynamics of linked populations in nature are a modification of [9, 10]. Appendix II reviews several alternative models.

If we assume that a change in time equals a change in age, then, as in Gutierrez and Wang (1976) and Wang et al. (1977), we can use a series of linked Leslie (1947) models or their continuous counterparts (McKendrick 1926; von Foerster 1959). An alternative continuous model for cases where we wish to describe the numbers and mass dynamics is that proposed by Sinko and

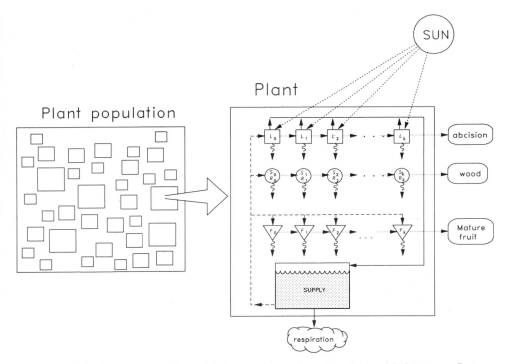

Figure 10.3. *An age-structured model of a population of plants and a model of the mass flow within one of them. Note that separate models are required for each subunit population and that they are linked via the metabolic pool model with allocation priorities indicated by subunit level.*

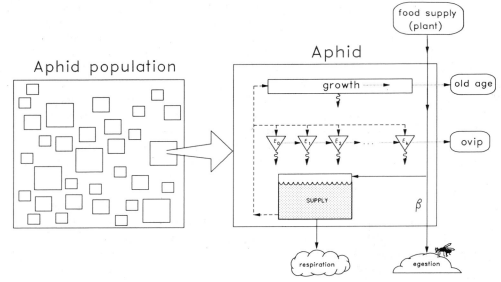

Figure 10.4. *An age-structured model of an herbivore (e.g., an aphid) population with the within individual mass flow dynamics illustrated.*

Streifer (1967) (see also Streifer 1974). As noted earlier, a distributed maturation time model would be preferred as it captures the variation observed in developmental rates (Manetsch 1976; Vansickle 1977; see Chapter 9). However, for simplicity, a system of coupled McKendrick–von Foerster and/or Sinko–Streifer models is used here to illustrate the general modular structure of plant and animal populations (e.g., Gutierrez and Wang 1977; Gutierrez et al. 1977a,b, 1984, 1988a,b, 1991a,b). In practice, these models would be discretized and programmed on a computer to evaluate the dynamics numerically (see, Wang et al. 1977, and/or Appendix II for discretization schemes).

A Plant Model

Assumes that each plant in a population has populations of age-structured subunits developing within it, [11]. The number dynamics model of whole plants is [11a], the models for leaf, stem, and root mass dynamics within the canopy of plants are [11b–d], respectively. The model for fruits including both number and mass dynamics is [11e]. Notice that we can mix number and mass models as needed.

Whole plant numbers

$$\frac{\partial \rho}{\partial t} + \frac{\partial \rho}{\partial a} = -\mu_\rho \rho(t, a) \qquad \begin{cases} \rho(t, 0) = B_\rho(t) \\ \rho(0, a) = \alpha_\rho(a) \end{cases}$$

$$[11a]$$

Leaf mass,

$$\frac{\partial L}{\partial t} + \frac{\partial L}{\partial a} = -\mu_L L(t, a) \qquad \begin{cases} L(t, 0) = g_L(t) \\ L(0, a) = \alpha_L(a) \end{cases}$$

[11b]

Stem mass,

$$\frac{\partial S}{\partial t} + \frac{\partial S}{\partial a} = -\mu_s S(t, a) \qquad \begin{cases} S(t, 0) = g_s(t) \\ S(0, a) = \alpha_s(a) \end{cases}$$

[11c]

Root mass

$$\frac{\partial R}{\partial t} + \frac{\partial R}{\partial a} = -\mu_R R(t, a) \qquad \begin{cases} R(t, 0) = g_R(t) \\ R(0, a) = \alpha_R(a) \end{cases}$$

[11d]

and Fruit numbers and mass

$$\frac{\partial F}{\partial t} + \frac{\partial F}{\partial a} + \frac{\partial}{\partial m} g_F(t, a, m)F = -\mu_F F(t, a, m) \qquad \begin{cases} F(t, 0, a) = B_F(t, m) \\ F(0, a, m) = \alpha_F(a, m) \end{cases}$$

[11e]

$\rho(t, a)$ is the number density function of plants of age a at time t, and similarly, $L(t, a)$, $S(t, a)$, and $R(t, a)$ are the age-structured mass density functions for leaves, stems, and roots, respectively. For fruits, $F(t, a, m)$ is the number density function for the population with attributes age a and mass m at time t. Mature seeds germinate and enter the plant population as new plants of age $a = 0$ with initial conditions for each of the subunit populations they contain (see below). The net death rates (i.e., birth–death rate) for whole plants, and the leaves, stems, roots, and fruits they contain are μ_ρ, μ_L, μ_S, μ_R, and μ_F, respectively. These net rates include the effects of shortfalls of photosynthate and mortality due to abiotic factors, herbivores, and diseases. When whole plants die, plant subunits within them die. The mass dynamics of the plant part subunit models depend on subunit age and the availability of photosynthate for allocation to them. α_ρ, α_L, α_S, α_R, and α_F in this model are the initial age distributions in the appropriate units, and g_L, g_S, g_R, and g_F are mass growth rate functions of the various plant subunits. B_ρ is the birth rate of new plants (accrued by any reproductive method) and B_F is the maximum initiation rate of new fruiting buds of initial per capita mass α_F. Both B_ρ and B_F may be heavily influenced by the photosynthate supply/demand ratio. These functions vary with species (or variety).

The mass of the plant is the sum of subcomponent populations. The dynamics of leaf mass but not numbers is modeled, but both could be accommodated as done for fruits. Developmental times and aging rates of the different sub-populations often differ, but this does not create undue problems as long as they are in the same units. Plants compete for resources such as light, water, and nutrients, and the subpopulations within them compete for photosynthate. Fruit always have a higher priority than vegetative growth. As we shall see, herbivores that tap the vascular system compete with plant subunits for resources.

A General Animal Model

A similar model [12] may be used to describe the dynamics of animal populations [$N(t, a, m)$] at any trophic level and of the developing reproductive subunits [$E(t, a, m)$] they might contain. The outflow from the reproductive submodel enters the parent population model as births producing overlapping generations (Fig. 10.4).

Parent number and mass dynamics model.

$$\frac{\partial N}{\partial t} + \frac{\partial N}{\partial a} + \frac{\partial}{\partial m} g_N(t, a, m)N = -\mu_N N(t, a, m) \qquad \begin{cases} N(t, 0, a) = B_N(t, m) \\ N(0, m, a) = \alpha_N(a, m) \end{cases}$$

$$[12a]$$

Embryo number and mass dynamics model.

$$\frac{\partial E}{\partial t} + \frac{\partial E}{\partial a} + \frac{\partial}{\partial m} g_E(t, a, m)E = -\mu_E E(t, a, m) \qquad \begin{cases} E(t, 0, a) = B_E(t, m) \\ E(0, m, a) = \alpha_E(a, m) \end{cases}$$

$$[12b]$$

Note that linked Sinko–Streifer models having attributes of age and mass are used to model the dynamics of both parents and the reproductive subunits they contain. The initial conditions for the parent organisms and the subunits they contain are α_N and α_E, respectively, B_N is the birth rate of new individuals, B_E is the initiation rate of new reproductive subunits, and g_N and g_E are the age-specific mass growth rates of whole individuals and reproductive subunits, respectively. Note that $B_E(t, m)$ is heavily influenced by food supply–demand considerations and $B_N(t, 0, m) = E(t, a_{max}, m)$ is the birth rate of new individuals of age (a_{max}) of mass m that flow from the reproductive submodel [12b] into the parent population dynamics model [12a]. The functions μ_N and μ_E are the net death rates for whole animals and the reproductive units they contain.

SOME EXAMPLES

Numerous models of different plant and animal systems incorporating mass (energy) flow and the rudiments of physiology and other aspects of biology have been written by myself and collaborators (e.g., Gutierrez and Wang 1976; Gutierrez et al. 1975, 1977a,b, 1984a,b, 1987, 1988a–c, 1990, 1994; Wang et al. 1977; Baumgärtner et al. 1986a,b; Graf et al. 1990a,b; Wermelinger et al. 1991a,b; Tamó and Baumgärtner 1993). *Some of the details may differ, but the underpinning model is the same.*

For simplicity, let us first assume that we are modeling the average plant (i.e., $\rho(t, a) = 1$ in [11a]), which multiplied by the total number of plants per area is a canopy model of equal aged plants (i.e., a crop). The system of equations describing this reduced model is [11b–11e]. Below, under the topic of objected oriented models, the full complexity of the dynamics of a population of plants of arbitrary age and mass structure is examined (see Fig. 10.3).

Cotton

Observed and simulated data for a cotton crop grown at Londrina, PR, Brazil are shown in Figure 10.5 plotted on a physiological time scale. Also shown are the weather data experienced by the crop. Solar radiation varied consider-

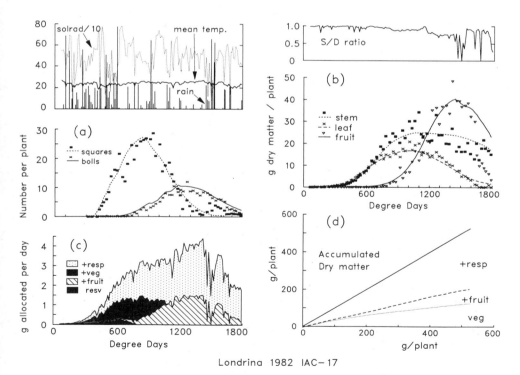

Londrina 1982 IAC–17

Figure 10.5. *Observed and predicted energy allocation in cotton* (Gossypium hirsutum *L. var IAC-17): (a) fruit dynamics; (b) dry matter allocation to leaves, stem plus root, and fruit; (c) age-specific energy allocation patterns; and (d) cumulative.*

ably over the season and together with water, soil nutrients, and temperature were important determinants of the photosynthetic rate. When light levels fell, per unit leaf photosynthetic rates fell (i.e., the supply), and when temperatures increased, per unit respiration rates and biomass demands increased. These are but two of the effects that fallout naturally in the model and may affect the dynamics of growth and development.

After germination, cotton plants grows vegetatively (Fig. 10.5b), and begin to produce fruit buds at some time t^*. Each fruit grows rapidly after flowering (Fig. 10.5a), at which time photosynthate is diverted to them (Fig. 10.5b) causing vegetative growth to cease and more than one-half of the fruit initiated to be shed. The fruits shed are usually those that have had the least amount of energy and time invested in them, but if stress is very severe, older fruits may also be shed. Leaf mass declines due to leaf abscission at maximum age and the reduced growth of existing leaves and slowed initiation of new leaves. Stem and root mass decline due to natural pruning. The shift from vegetative to reproductive growth may be sudden and complete as seen here, or may be partial when fruit demands are low due to past depletion of fruits by herbivores (or due to low intrinsic fruit growth rates in some species or varieties). Cassava plants, for example, produce small fruits relative to their size, and often sustain simultaneous vegetative and reproduction growth. Finally, note that the cumulative allocation to different processes is constant (Fig. 10.5d), this despite the irregular pattern of daily photosynthesis (Fig. 10.5c).

Arthropods

The daily per capita age-specific consumption and allocation rates in the viviparous pea aphid to respiration, egestion, and assimilation to growth and reproduction were shown in Figure 6.8a. As in cotton, the partitioning ratios to egestion, respiration, and assimilation remained relatively constant (0.176, 0.338, and 0.486 respectively). The rate of assimilation to reproduction is approximately the same as that assimilated to maximum growth at the time of the adult molt. At this time, many embryos within the female begin exponential growth causing the supply to become less than the demand, the youngest embryos to be aborted, and the growth rates of those retained to slow down.

The biology and patterns of reproduction in the female lady beetle *Hippodamia convergens* (G-M) (Fig. 10.6) is very different from the pea aphid, but the same model reproduces its per capita dynamics. With unlimited prey, dry matter allocation to egg production in adult females continues at a constant rate until late in the beetle's life (Fig. 10.6b and c, Gutierrez et al. 1987). This occurs because the demands of the developing ova are small relative to the ability of the females to consume and assimilate food. If food is in short supply, reproduction may slow or stop (see Chapter 6). Aphids and coccinellids produce extreme patterns, but more extreme examples as well as various intermediates occur in both plants and animals.

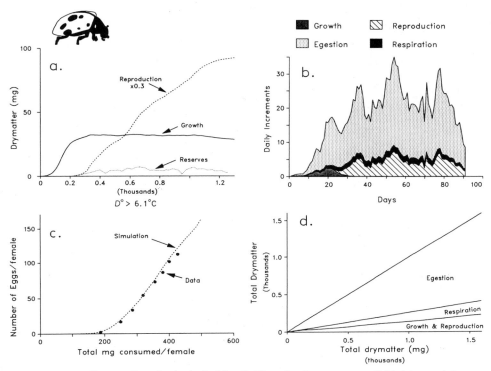

Figure 10.6. *Energy allocation in the ladybeetle* Hippodamia convergens *G-M: (a) cumulative age-specific energy allocation, (b) daily allocation, (c) comparison of predicted and observed cumulative oviposition, and (d) cumulative allocation. [Reprinted with permission from Gutierrez et al. 1987.]*

Plant–Herbivore Interactions

The linkages between a plant and a sap feeding herbivore (say an aphid) are depicted in Figure 10.7. A population of plants is shown with one of them being attacked by a population of aphids. Each plant may have a different infestation level that taps its vascular system and competes with the plant's subunits for resources. Of course, other herbivores may attack leaves, stems, roots, or fruits, and each can be modeled just as easily. The interspecific competition between herbivores is indirect via effects on plant growth, while intraspecific competition for available resources is direct. Herbivores attack either the supply (e.g., aphids or defoliators) or the demand (e.g., fruitivores) side of the system.

Herbivores such as boll weevil (*Anthonomus grandis* Boh.) attack cotton buds. This attack affects future patterns of fruit demands for photosynthate. The boll weevil lays its eggs in cotton fruits where its larvae develop—in a

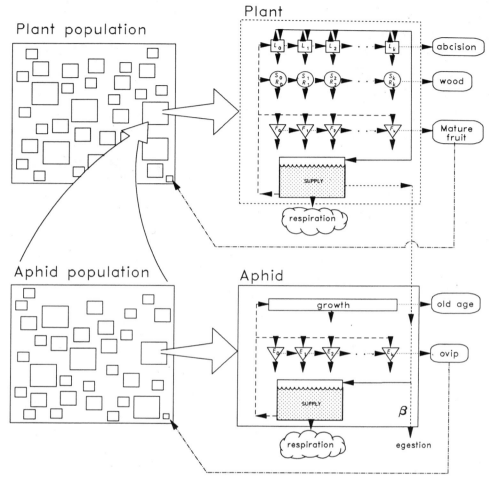

Figure 10.7. *A conceptual model of the intra- and interspecific interaction of plants and herbivores within a plant using energy flow rates.*

manner akin to a parasitoid. The photosynthetic rate is affected by the loss of fruits, and hence is allocated primarily to vegetative growth. In heavily defruited plants, biomass accumulation to stem tissues may be three times higher than in unattacked plants (Fig. 10.5 vs. 10.8). The production of new buds is part of vegetative growth and continues unabated (Fig. 10.8a). This finding is not new, as most gardeners know that one can prolong flower production by harvesting them before the seeds begin growing. One might be tempted to invoke the complexity of hormonal control of energy allocation in plants to explain this, but in our model, the supply/demand ratio controls all rates pro-

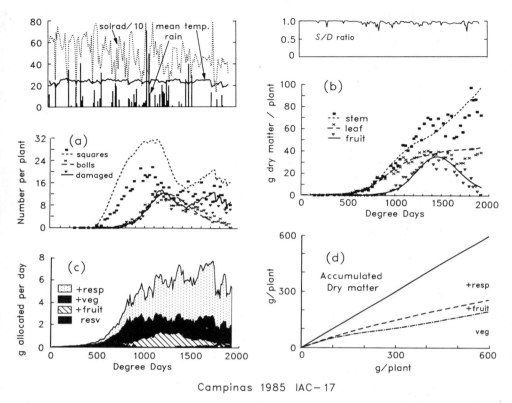

Campinas 1985 IAC-17

Figure 10.8. Observed and predicted energy allocation in cotton (Gossypium hirsutum L.) with boll weevil and high-soil fertility: (a) fruit dynamics; (b) dry matter allocation to leaves, stem plus root, and fruit; (c) age-specific energy allocation patterns; and (d) cumulative allocation. [Gutierrez et al. 1991a.]

viding sufficient detail to predict the observed plant growth dynamics. (Note that the model is independent of the data it reproduces.)

Another obvious point that needs to be demonstrated is that soil nutrient levels have profound effects on plant—herbivore (and higher trophic level) interactions. Weevils affect the demand side of the S/D ratio and poor plant nutrition affects the supply side. Plant and weevil dynamics are controlled by their respective supply/demand ratio. Thus, plants growing on poor soil have lower photosynthetic rates and, hence, slower growth and fruit bud production rates, and higher fruit shed rates. The effects of these interactions are seen by comparing the plant growth dynamics in Figures 10.5, 10.8, and 10.9. In the pest-free crop, observed weather and nonlimiting edaphic factors determine how much of the plant's genetic potential to produce seed is expressed in the absence of herbivores (Fig. 10.5). Figure 10.8 shows the added effects of boll

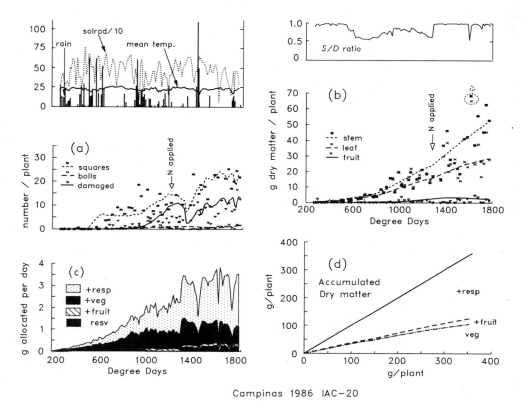

Campinas 1986 IAC−20

Figure 10.9. *Observed and predicted energy allocation in cotton* (Gossypium hirsutum *L.): with boll weevil and low soil fertility: (a) fruit dynamics; (b) dry matter allocation to leaves, stem plus root, and fruit; (c) age-specific energy allocation patterns; and (d) cumulative allocation. [Gutierrez et al. 1991a.]*

weevil predation of fruits within plant dry matter allocation in a crop with high-soil nutrients, and Figure 10.9 has the added effects of poor plant nutrition.

This model has practical as well as theoretical applications. Below, we use the cotton model to examine economic issues in cotton production and protection, and the economics of fitness and adaptedness of sylvan cotton and the cotton boll weevil in an ecosystem context.

Some Economic Considerations. There is always the danger in using models that we might be tempted to find explanations beyond what is reasonable. Despite this caution, realistic models such as this have become invaluable tools in applied ecology for the evaluation of plant–herbivore interactions and to estimate the economic impact of pests in agricultural crops for the development of pest management strategies. Realistic models of crop–pest–natural enemy systems allow examination of the interaction from several perspectives.

Successful applications of the model to cotton crop production and IPM were recently summarized by Gutierrez (1995) and include: a general IPM strategy for fruitivores in cotton based on the ratio of cumulative number of buds initiated to the cumulative number damaged (Gutierrez et al. 1981b, Gutierrez and Daxl 1984); estimates of the efficacy of sex pheromones for pink bollworm control (Stone and Gutierrez 1986); evaluation of the economics of alternative control strategies for the pathogen verticillium wilt (Regev et al. 1990); development of estimates of the effects of defoliators and sap sucking pests on yield; estimation of the economics of pests that kill whole plants (Dos Santos et al. 1989); and evaluation of the role of complex parasitoid biologies on whitefly control and crop yield (Mills and Gutierrez, unpublished).

Problems involving multiple pests and trophic levels may appear vexing, but a realistic model provides a mechanism to dissect the problem into its component parts. Starting with the notion that each pest affects either the supply or demand side of the S/D ratio, the model allows simulation of the simultaneous interaction of all their effects. The modular structure of the model enables us to examine different combinations with simple true–false instructions that add or deplete species or factors. The computing time of the model is approximately 15 seconds, on a desktop computer for a season that may include several interacting pests, making possible the evaluation of hundreds of scenarios that could not be done in the field. With the use of marginal analysis (or other techniques), we can assess the effects of each pest and their interactions on various aspects of plant growth development and yield. As an example, the plant bug *Lygus hesperus* Knight was once considered the key pest of cotton in the San Joaquin Valley of California. Our model explained why this was highly improbable resulting in millions of dollars annually in savings in pesticide applications (see Gutierrez 1995 for a review).

The determination of optimal solutions for multipest problems is considerably more difficult. Consider the simple problem of determining the optimal pattern of spraying for boll weevil control in cotton. Assume that there are 10 time periods in the season, and we merely wish to ask whether we should or should not spray in each of them. This problem results in a *Shoemaker's combination*[1] of options (i.e., $2^{10} = 1024$). Of course, if we consider multiple pesticides, different combinations and dosages, as well as multiple pests, and so on, the problem becomes very large and not amenable to brute force methods of simulation and global search—it is a problem of dimensionality. Furthermore, reducing the problem and solving it by various optimization techniques is extremely difficult. Appendix IV outlines some of the optimization methods that have been used to evaluate such complex problems.

[1]This example was named after Christine Shoemaker who repeatedly used it to illustrate this very important point.

The Economics of Fitness: Sylvan Cotton and the Cotton Boll Weevil

We explore the question of the coevolved reproductive strategies of cotton and its herbivore the cotton boll weevil. This herbivore is thought to have evolved on *Hampea* and later moved to cotton (Burke 1976). Using a model of sylvan cotton and the weevil we can ask what would happen to joint fitness if their reproductive rates increased or decreased, given that the remaining biology is unchanged. It would be impossible to examine this question in nature as the biology of a species is largely fixed except in evolutionary time, but this question can be examined by using our model. The question is largely one of the evolution of resource allocation strategies that enhance long run persistence and adaptedness (Gutierrez and Daxl 1984). To examine the question, Why do species allocate energy (i.e., behave) as they do (cf. Fisher 1930)? requires that the question be asked in an ecological context (Gilbert et al. 1976) and evaluated near equilibrium. Models that include the realism of resource acquisition and allocation and are driven by weather and edaphic factors provide a reasonable tool for examining such problems.

For example, Gutierrez and Regev (1983) made numerous simulation runs of the model varying either the fruit bud production rate of cotton (R_c) or the oviposition rate (R_w) of the weevil.[2] (Note that R_c and R_w are multiples of the observed rates.) The numbers of mature cotton fruits (seeds, F_c) at the end of the different runs are plotted on R_c across different levels of R_w (Fig. 10.10a). Likewise, the season long productions of diapause weevil adults (F_w) are plotted on R_w across different levels of R_c (Fig. 10.10b). These curves can be viewed as yield effort functions, and F_c and F_w are measures of fitness.

The Plant's Point of View.   In the absence of the weevil ($R_w = 0$), the highest seed yields F_c accrue at values of R_c higher or lower than the observed value. The observed value is in a trough. However, when the weevil is included and its fecundity increased, seed yields decline. Interestingly, the functions for F_c now peak near ($R_c = 1$, $R_w = 1$) and decline for smaller and larger values of R_c. The interpretation of this is that cotton invests energy in excess fruiting capacity to accommodate anticipated boll weevil predation (i.e., the difference between points A and B in Figure 10.10a).

The Weevil's Point of View: Similar plots of F_w on R_w across different levels of R_c are shown in Figure 10.10b. The broad plateau of the function indicates that F_w is little affected across reasonable changes in R_w. In contrast to cotton, maxima for F_w in the weevil shift to higher R_w values as R_c increases, suggesting a strong bottom-up influence.

Coevolution: The equilibrium properties of the system may be examined graphically by first assuming equilibrium fitness for F_C^* and F_w^* and a constant

---

[2]Initial weevil population level observed in a 1972 experiment in Nicaragua was arbitrarily selected (see Gutierrez and Daxl 1984).

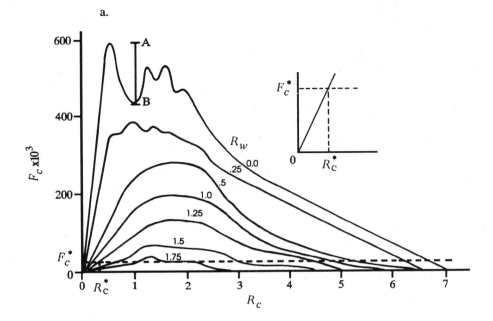

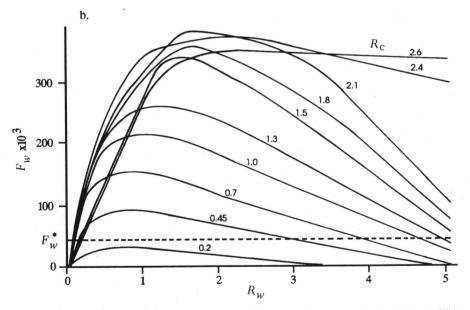

Figure 10.10. Plots of fitness (F_c, F_w, respectively) for cotton (a) and the cotton boll weevil (b) from numerous simulation studies runs across a wide range of fruit bud production rates (R_c) and of weevil's fecundity (R_w) [cf. Gutierrez and Regev 1983].

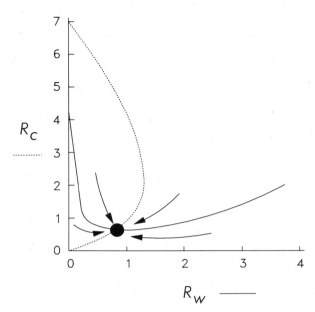

Figure 10.11. *A summary of the graphical analysis of the cotton boll weevil system: plots of equilibrium R_c^* values on equilibrium values of R_w^* from both the plant's (solid line) and weevil's point of view (dashed line) [cf. Gutierrez and Regev 1983].*

survival rate. Equilibrium occurs at F_C^* when $[F_c(R_c, R_w, y + 1)/F_c(R_c, R_w, y)]$ = 1 between seasons y and $y + 1$. Similar arguments are made for F_w^*. The critical equilibrium levels are represented by the dashed horizontal lines in Figures 10.10a and b. Possible equilibrium occurs at the intersections of the function and the equilibrium level, and each has coordinates R_c and R_w [i.e., $F_c^*(R_c, R_w)$ for the plant and $F_w^*(R_c, R_w)$ for the weevil] (see inset in Fig. 10.10a, Peterman et al. 1979). The complete set of equilibrium points falls on the lines plotted in Figure 10.11 (i.e., – – –, for the plant; ——— for the weevil).

The sylvan cotton weevil system has only one common equilibrium point (R_c^*, R_w^*) and it is reasonably near the observed values $R_c = 1$ and $R_w = 1$. This suggests a continuing tension of evolution toward the observed reproductive level given the current abiotic conditions under which the two species coevolved. The joint solution is constrained maximized fitness for both, suggesting that the observed strategy is best given weather and the uncertainty of weevil attack from season to season. The plant has one and possibly two seeding periods during the season, while the weevil has a generation every 15 days (i.e., 7–12 generations). The model is entirely deterministic and qualitative, and provides no information as to the time course of the evolution.

OBJECT ORIENTED MODELS

Judson (1994) asserts that we are witnessing a major shift away from models governed by general equations and towards those that create each and every individual in the ecosystem in order to generate the dynamics of the system as a whole. She classifies such models as *individual based distribution models* that lump individuals together according to common characteristics or *individual-based configuration* models that track each individual in the population. Models of individuals (objects) may provide a suitable framework for integrating the interactions between members of a population, and across trophic levels. This approach was first proposed, *albeit* using other terminology, by Holling (1966) who studied the component of arthropod predation. He was at that time constrained by a lack of cheap, fast computing power, and possibly by the lack of relevant computer programs to facilitate his task. Recently, Huston et al. (1988) emphasized the utility of modeling individual members of a population to integrate the different levels of organization of the traditional ecological hierarchy—physiological ecology, behavioral ecology, autoecology, population ecology, community ecology, and ecosystem ecology.

The activities of individuals (or group of similar types) and their interactions are being used to examine total population responses. For example, Jones (1981), Jones et al (1975), Casas (1989), and others have used individual behavior to study population attributes in insects, but not to model population consequences. Sequeira et al. (1991) used object oriented simulation to model plant growth and organ–organ interactions, but the level of detail required was large, and hence beyond the scope of this expose. The recently edited book by DeAngelis and Gross (1992) on individually based models is an important starting point for this approach.

A Simple System

Even aged plants spaced at regular intervals would, in the absence of density dependent and other limiting effects, grow to be the same size (Fig. 10.12a). In nature, plants have different ages and genetical make up, experience different initial conditions of resources such as space and essential nutrients, and these differences are compounded over time causing some individuals to thrive and others to be suppressed (i.e., Fig. 10.12b). These interactions between plants are implied in Figures 10.3 and 10.7 and [11]. The transition from a model of individual growth and development to that of a population of individuals with different attributes and developmental history is relatively straight-forward. Let us first take a simpler approach.

If $M[t, a, (i, j)]$ is the mass of an individual plant (or animal) at time t of age a at coordinates (i, j), then a metabolic pool model of its growth might be

$$M(t + \Delta t, a + \Delta a, i, j)$$
$$= M(t, a, i, j) + \{\beta(i, j)S^*(t, i, j) - z(t, i, j)M(t, a, i, j)\}\lambda(i, j) \quad [13]$$

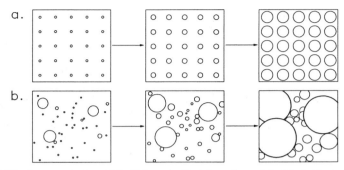

Figure 10.12. *A conceptual model of intra specific interactions: (a) all individuals begin life at the same time and no competitive advantage accrues over time, and (b) individuals begin life at a different time and competitive advantages accrue over time. [Reprinted with permission from Huston et al. 1988.]*

where $\beta(t, i, j)$, $z(i, j)$, and $\lambda(i, j)$ are the physiological parameters of the metabolic pool model. These parameters may be different for each individual. In this model, competition occurs via factors that affect the capacity to acquire and assimilate resources (S^*). If our model assumes that the mass of an individual is an indicator of its ability to obtain resources relative to its nearest neighbors, this would produce patterns of individual growth similar to those shown in Figure 10.12b.

Modeling herbivores or higher trophic level individuals as objects might not make sense, especially if the individuals are highly mobile, their per capita effects are not large, or the computation time is not practical. A compromise might be to model individual plants with herbivore and higher trophic level organisms treated as populations on individual plants (Fig. 10.7). Each plant could be of a different genotype or species with different levels of resistance to the herbivores, and grow on soils with different levels of nutrition and water holding capacity. Qualitative and quantitative differences develop among plants as they grow and compete, and this often stimulates different rates of herbivore attraction, growth and reproduction, aggregation, and other responses. The injury caused by herbivores in turn influences the capacity of the plants to grow, reproduce, and compete, ultimately affecting the herbivore growth and patterns of population dispersion. Herbivores affect the supply or demand side in each plant differently and, hence, the dynamics of plant growth and reproduction, with a lag in their own dynamics on the plant. All of these interactions are implied in Figures 10.7 and 10.12b and are captured by the models (e.g., [11] and [12]). Below we compare canopy and object-oriented models of the tri-trophic cassava system.

The Tri-Trophic Cassava (*Manihot esculenta* Crantz) System of West Africa

A Canopy Model: Predicting the efficacy of natural enemies prior to their introduction for biological control of pests has proven difficult (Huffaker et al.

1971; Gutierrez et al. 1984, 1990, 1993, 1994; Godfray and Waage 1991). Most tri-trophic studies have been theoretical in nature, but my colleagues and I have stressed biological realism (see also Neuenschwander et al. 1986, 1987; Neuenschwander and Madojemu 1986; Pijld et al. 1990; Mills and Gutierrez unpublished). Recently, the roles of two exotic encyrtid parasitoid [*Epidinocarsis lopezi* and *Epidinocarsis diversicornis* (Howard)], some native natural coccinellid predators, a fungal pathogen (*Neozyites fumosa*), and weather and edaphic factor effects on the dynamics of the cassava mealybug *Phenacoccus manihoti* in West Africa were examined (Gutierrez et al. 1988a,b, 1993, 1994). The modular structure of the model enabled us to add and delete species with ease to examine the effects on the interactions.

The predictions of the canopy model of the cassava system compared well to the independent field data (Fig. 10.13a, c). Among the important findings were (a) the functional and numerical responses of either parasitoids alone or in combination were insufficient to explain the observed dynamics of the mealybug; (b) rainfall and its enhancement of the fungal pathogen suppress CM

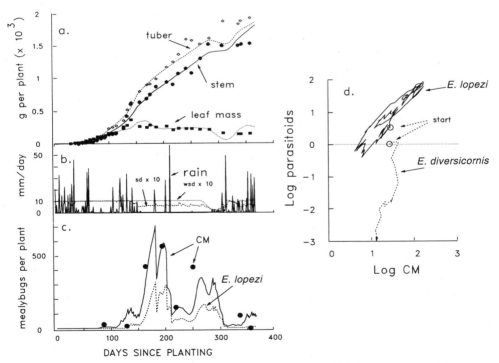

Figure 10.13. Simulation of the interaction of (a) cassava and (c) the cassava mealybug as affected by the parasitoid Epidinocarsis lopezi and native coccinellid predators and weather. The temperature and rain patterns experienced by the organisms are shown in (b), and the phase plane dynamics of log parasitoids (E. lopezi and E. diversicornis) on log mealybug numbers are shown in (d) [cf. Gutierrez et al. 1993].

numbers sufficiently during the wet season so that the parasitoid *E. lopezi* with its efficient search could regulate mealybug densities at low levels during the dry season; (c) weather (Fig. 10.13b) and soil factors (e.g., nitrogen and water) affect plant growth rates directly, and mealybug size and number dynamics and parasitoid sex ratios indirectly; (d) the effects of host size dynamics on sex ratios favor *E. lopezi* over *E. diversicornis*; (e) the importance of low rates of parasitoid immigration for the successful regulation of CM was demonstrated; (f) the fivefold higher host finding capacity of *E. lopezi* enhances its dominance over *E. diversicornis*; but (g) it is the sway of *E. lopezi* in cases of multiple parasitism that ultimately causes the competitive displacement of *E. diversicornis* from the system during periods when few hosts are available and/or when weather induced plant stress decreases host size favoring a stronger female biased sex ratio in *E. lopezi* than in *E. diversicornis* (the phase diagram, Fig. 10.13d).

A Mixed Object-Oriented Model: As indicated above, modeling herbivores and higher trophic levels as objects might not make sense for several reasons.

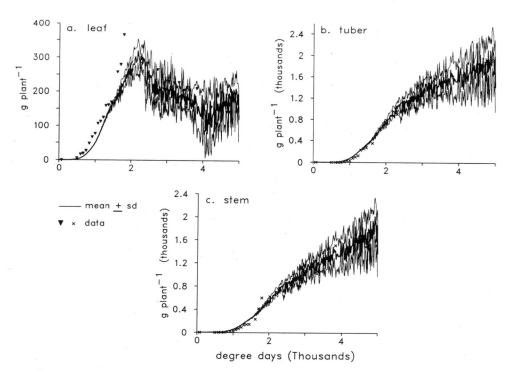

Figure 10.14. *An object oriented simulation of cassava plant model with populations of cassava mealybugs and parasitoids within: The simulated leaf (a), tuber (b), stem (c) dynamics showing estimated means and standard deviations (see Fig. 10.13 for deterministic results), and (d) the within field distribution of soil moisture and nutrient levels.*

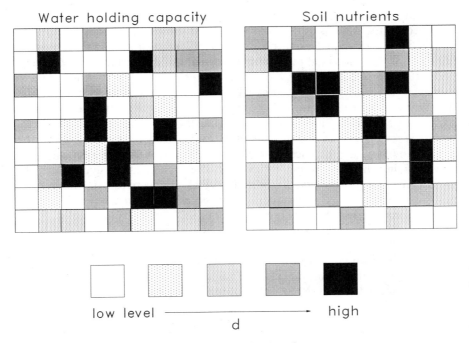

Figure 10.14. (Continued)

We simulated individual randomly spaced cassava plants each having populations of the mealybug and its two parasitoids and coccinellid predators. Each was assumed to grow in soil with unique levels of nutrition and water holding capacity (Fig. 13.14d), and the movement of pests and natural enemies between plants and the times of initial infestation were random.

The same levels of detail as seen in Figure 10.13 for the deterministic canopy plant and animal populations were obtained from the object oriented stochastic model (Fig. 10.14a–c). In the latter, daily means and variances of leaf, stem, and tuber mass accumulation were estimated from a sample of five plants drawn at random from the simulated population. The means gave patterns similar to those produced by canopy plant models but they were not smooth. The dynamics of the mealybug and parasitoid populations were not radically different from those shown in Figure 10.13, and hence they are not illustrated.

EPILOGUE

Analogous biological processes occur at all trophic levels, and these have been exploited in the development of realistic age- and mass-structured multitrophic population dynamics models. At the per capita level, the acquisition (functional response), and allocation of resources (assimilation) are the same and only the units and the biology described differ (Chapters 5 and 6), but all may be described by the same physiological model. At the population level, the individuals may have populations of subunits within them, but the population dynamics of the parent and their subunit populations are described by the same model (Gutierrez and Wang 1976). These analogies and symmetries allowed the development of general theoretical models (Chapters 7 and 8) as well as realistic models of single species (Chapter 9) that were combined here with relative ease in a multitrophic setting. The development of object oriented models is reviewed, but no conclusions as to their superiority are made.

These models have practical use in pest control, as well as theoretical utility for examining the coevolution of species. The models allow questions to be asked concerning the interactions from several perspectives, and because of their modular structure, allow questions to be asked in an ecosystem context by examining different combinations of factors and/or interacting species. These properties proved useful in the evaluation of the effects of abiotic and biotic factors on the coevolution of a plant and herbivore, and for evaluating the potential biological control of cassava mealybug and whiteflies. However, despite their broad utility, all models, including ours, will always be incomplete, and hence must be considered only as guides to understanding complicated biological interactions.

11

Regional Dynamics

. . .beware of computer driven disasters.
To err is human, to really screw up requires a computer.
—Uri Regev

Populations of crickets, locusts, aphids, leafhoppers, armyworms, mice, lemmings, red tides, fleas, and other species may reach levels of biblical proportions over large geographic areas. Such outbreaks may go unnoticed in sparsely populated areas until public interests (e.g., agricultural lands, forests, and fisheries) are threatened. In general, forecasting such outbreaks on a regional basis is desireable if we are to manage the gobal ecosystem in a sustainable manner.

Populations of species are often discontinuous in time and space due to the action of biotic and abiotic factors. Species and some of their populations may have distinct sets of abiotic requirements for growth and development (i.e., the *Fundamental Niche*, cf., Hutchinson, 1959; see Chapter 2). Many species may be regulated by natural enemies and competitors (Hutchinson's realized niche), but the abundance of others may be largely determined by the stochasticity of abiotic factors such as weather.

Traditional population models describe changes in population size over time as a function of changes in resources, natural enemies, and competitors. However, for regional analyses, it may not be possible (or desirable) to model the numerical dynamics of species over a large geographic area except in some qualitative way, because the initial conditions and the nature of the biotic and abiotic factors affecting them are usually not known. This may not matter as it may be sufficient to evaluate the potential for the outbreak of a species. For this reason, the focus in this chapter is on the effects of abiotic factors on the potential for population growth. This problem in various forms has a long history in physiological ecology (see Andrewartha and Birch 1954).

A CONCEPTUAL MODEL

Akçakaya et al. (1988) proposed that the simplest model of intermediate complexity which explains the dynamical properties of a population is a second-order model that describes the population dynamics as a function of physiological variables that depend on the resource level. For example, the potential growth rate of a population (dN/dt) may be described as a first-order function of all important abiotic variables,

$$\frac{dN}{dt} = F(x_1, x_2, \cdots, x_n) \quad \text{for} \quad i = 1, 2, \ldots, n \quad [1]$$

If one could formulate this model appropriately, one could examine the second-order effect of each factor x_i by taking the derivative of $F(\cdot)$ with respect to each factor (i.e., dF/dx_i). This procedure would enable us to determine whether a level of a factor has positive or negative effect on the population growth rate.

Formulating the function $F(\cdot)$ in a realistic manner is difficult but tractable. Simulation models (e.g., Gutierrez et al. 1988a,b and c) are increasingly common in the literature, but they are rarely amenable to mathematical analyses of this kind. Here we take a simpler approach and decompose the problem into its components by examining the effects on the per capita growth rate of different factors singly and then combining their effects.

PHYSIOLOGICAL EFFECTS

We know that under nonlimiting abiotic conditions, the developmental rate of poikilotherms has a sigmoidal response to temperature (τ) (see Chapter 3). This effect is first order. The second-order effect (i.e., $dR/d\tau$) is parabolic indicating minimum, optimum, and maximum temperatures for development. We also know that for per capita growth and reproductive rates, the intrinsic rates of population growth and mean longevity are parabolic functions of temperature with characteristic minimum, optimum, and maximum (see Chapters 4 and 6). It is often presumed that the first- and second-order responses of poikilotherm growth to levels of other essential resources have similar shapes (Figure 11.1). This suggests that a resource in short supply (von Liebig's law of the minimum 1840) or in excess may by itself limit a species. If several factors are not at optimum levels, all may affect the rates of growth and development. Our task is to estimate the combined effects of an arbitrary set of factors on the potential for population growth on a regional basis. Below we review three approaches that converge to similar qualitative models.

Growth Indices

Fitzpatrick and Nix (1970) developed empirical parabolic growth indices to predict the effects of temperature [$TI(t)$] and the availability of water [$WI(t)$],

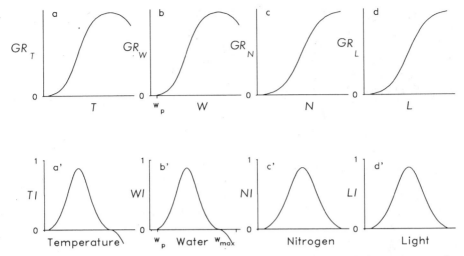

Figure 11.1. *Fitzpatrick and Nix (1970) growth indices: (a–d) hypothetical growth rates of a plant species with respect to different levels of a resource, and (a'–d') the normalized second-order effects or growth indices.*

nitrogen [$NI(t)$] and light [$LI(t)$] on the growth rate of temperate and tropical pastures across Australia. Each index is humped and each predicts a minimum, maximum and optimum level (Figure 11.1). this is an important point, as it implies that the level of a resource may be so high that its second-order effect on growth is negative (e.g., too much soil nitrogen slows growth). Fitzpatrick and Nix proposed that the effects of the factors were multiplicative rather than additive, hence the overall growth index [$GI(t)$] is

$$0 \leq GI(t) = TI(t)\, MI(t)\, NI(t)\, LI(t) \leq 1 \qquad [2]$$

If all but one factor are unity, then the growth of the population at time t will be regulated by the factor in short supply. However, if all indices are positive but less than unity, then growth at time t is slowed by the compounded effects of all partial shortfalls. We can also derive growth indices by computing growth rates under different levels of resources.

Growth Rates

Under resource level (R) and temperature τ, the per capita growth rate (GR) is the acquisition rate corrected for respiration [$r(\tau)$] and other costs (θ) (see Chapter 5).

$$GR(t) = \theta\left(1 - \exp\left[-\frac{\alpha R}{D(\tau)N(t)}\right]\right) D(\tau) - r(\tau) \qquad [3]$$

where N is the population density and is included in the model to incorporate the effects of intraspecific competition. The parameter $D(\tau)$ is the genetic maximum per capita demand rate for resource at τ, and α is the proportion of resource available for use during dt. If there are n other essential resources, their effects can be incorporated as the product of the supply/demand ratios ($\phi_i = S_i/D_i$) (see Chapter 6 Appendix 6.1).

$$GR(t) = \prod_{i=1}^{n} \phi_i \, \theta \left(1 - \exp\left[-\frac{\alpha R_o}{D(\tau)N(t)}\right]\right) D(\tau) - r(\tau) \qquad [4]$$

In this case, the $0 < \phi_i < 1$ are concave with respect to the level of resource and scale the growth rate. Across temperatures, [4] is parabolic, hence shortfalls of other nutrients serve to lower and narrow it from the right (see Chapter 8). If we define GR_{max} as the maximum growth rate under nonlimiting conditions, we can compute an overall growth index given the levels of other resources

$$GI(t, \cdot) = GR(t, \cdot)/GR_{max} < 1 \qquad [5]$$

Note that both GI and GR are parabolic with respect to *temperature* (Gutierrez et al. 1994). This approach does not accommodate excessive levels of a resource.

A Continuous-Time Markov Model Approach for Modeling Growth Rates

An interesting approach that yields a similar model is the continuous-time Markov model used by Olson et al. (1985) to characterize the effects of light intensity (ψ), water availability (W), and limiting nutrient (N) on biomass (M) accumulation rates (i.e., GR) in plants. No doubt other factors could be included. The mathematics are extensive, hence interested readers are referred to the original article.

Figure 11.2 illustrates the five states of a plant growth rate model. The coefficients α_i ($i = 1, \ldots, 8$) represent interfaces between the model system and the environment. The system can go from state 1 to state 2 or 5. To go to state 2, light energy (ψ) is required for photosynthesis, hence the transition rate between states 1 and 2 is proportional to ψ (i.e., $\lambda_{12} = \alpha_1\psi$). The system can also go to state 3 from 2 or return to 1 with transition rate λ_{21}. Likewise, the mean transition rate between states 2 and 3 is proportional to water availability (i.e., $\lambda_{23} = \alpha_2 W$). The transition from 3 can only be to 4 with the transition rate $\lambda_{34} = \alpha_2 N$, where N is some rate-limiting nutrient (e.g., nitrogen, potassium, or phosphorous, and so on). The transition from state 4 to state 1 is the biomass producing step with mean transition rate λ_{41}. The dashed line from this transition rate to the box labeled M containing the biomass of the plant including storage is information flow. The respiration demand is proportional

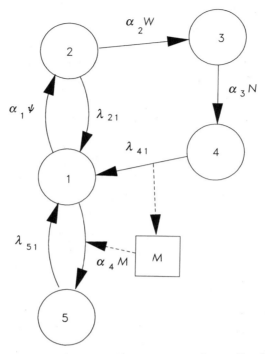

Figure 11.2. *The transition diagram for a three resource continuous-time Markov plant model. [Redrawn from Olson et al. 1985.] Note that material flow is indicated by the solid arrows and information flow by the dashed arrows.*

to M and is characterized by the mean transition rate $\lambda_{15} = \alpha_4 M$, and that from state 5 back to state 1 is λ_{51}.

Olson et al. (1985) argue that the plant's metabolic system is composed of millions of subunits, each of which may be in any state, and the steady-state probability π_i represents the fraction of the system units in state i at any time t. The transition matrix (Table 1) was formulated based on Figure 11.2, and is composed of all mean transition rates such that the element in row i column j equals λ_{ij}, those on the diagonal are the negative sums of all other row

TABLE 1. Transition Matrix Based on Figure 11.2

Stage	1	2	3	4	5
1	$-(\alpha_1\psi + \alpha_4 M)$	$\alpha_1\psi$	0	0	$\alpha_4 M$
2	λ_{21}	$-(\lambda_{21} + \alpha_2 W)$	$\alpha_2 W$	0	0
3	0	0	$-\alpha_3 N$	$\alpha_3 N$	0
4	λ_{41}	0	0	$-\lambda_{41}$	0
5	λ_{51}	0	0	0	$-\lambda_{51}$

elements, and the row elements sum to zero. The transition matrix Λ is left multiplied by the row vector $\boldsymbol{\pi}$ consisting of all π_i ($i = 1, 2, \ldots, 5$) and set equal to the null vector $\mathbf{0}$ at steady state (i.e., $\boldsymbol{\pi}\Lambda = 0$). Now, $\pi_1 + \pi_2 + \pi_3 + \pi_4 + \pi_5 = 1$ and must replace one of the equations in the system to yield a closed-form solution. The steady-state probabilities of being in each of the states are

$$\pi_1 = \frac{[\alpha_3 N \lambda_{41} \lambda_{51}(\lambda_{21} + \alpha_2 W)]}{S}$$

$$\pi_2 = \frac{[\alpha_1 \psi \alpha_3 N \lambda_{41} \lambda_{51}]}{S}$$

$$\pi_3 = \frac{[\alpha_1 \psi \alpha_2 W \lambda_{41} \lambda_{51}]}{S}$$

$$\pi_4 = \frac{[\alpha_1 \psi \alpha_2 W \alpha_3 N \lambda_{51}]}{S}$$

$$\pi_5 = \frac{[\alpha_3 N \alpha_4 M \lambda_{41}(\lambda_{21} + \alpha_2 W)]}{S}$$

where S is the normalized sum of the numerators of the π_i.

The relative growth rate (RGR) may be written as the difference between the steady-state biomass production rate [i.e., the mean transition rate λ_{41} times the steady-state probability of being in state 4 (i.e., π_4)] and the rate of respiration (i.e., the mean transition rate between 1 and 5 [$\alpha_4 M$ times the probability of being in state 1 (i.e., π_1)]

$$RGR = \frac{1}{M}\frac{dM}{dt} = \lambda_{41}\pi_4 - \alpha_4 M \pi_1 \qquad [6a]$$

Substituting identities and simplifying with constants e_1, e_2 and e_3, we get the decreasing function

$$RGR = \frac{1 - e_1 M}{e_2 + e_3 M} \qquad [6b]$$

Multiplying both sides by M we get the parabolic realized growth rate with the effects of other factors included

$$GR = \frac{dM}{dt} = \frac{M - e_1 M^2}{e_2 + e_3 M} \qquad [7]$$

Note that when both functions cross the x-axis, the biomass used for (dark) respiration exceeds that produced via photosynthesis. As before, growth indices

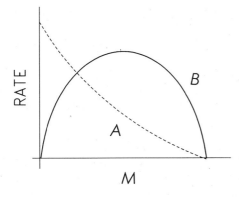

Figure 11.3. *Relative growth rate (A, – – –) and absolute growth rate (B, ——) are plotted on M. [Redrawn from Olson et al. 1985.]*

are obtained by normalizing *GR*. Both *RGR* and *GR* can be written as functions of ψ, *W*, and *N* singly or in combination, and allows one to characterize growing conditions for the plant in two dimensions (Fig. 11.3) and not four. If there is a shortcoming to this model, it is that the various transition rates must be estimated and are likely to be time varying.

IMPLEMENTATION

We could use any of the three models to forecast the potential for the growth and development of a species, but [2] and [5] are most appropriate for use on a regional scale. Both require weather and models of the dynamics of edaphic factors to drive them (see Gutierrez et al. 1974b). Unfortunately, a major unresolved problem for sparsely populated areas or areas with poor technical infrastructure is the lack of relevant real-time weather and edaphic data. How to implement the biological and physical models for regional analyses and how to summarize the data are additional problems. Satellite imagery for remote sensing of abiotic factors over wide geographic areas and geographic information technology (GIS) to organize data to run models and to summarize model output are important developing technologies (Cherlet and DiGregorio 1991).

Satellite Imagery

Daily weather data tends to be sparse over vast geographic areas of the planet such as Africa, and considerable effort is in progress to correct this deficiency using remote sensing satellite technology. Many different kinds of data are already available, and collections of other kinds are being planned as components of *mission to planet earth* (Matson and Ustin 1991). Satellites provide spatial and spectral information on a synoptic scale that will be increasingly useful in ecological research (Roughgarden et al. 1991). Wickland (1991) provides a tutorial on remote sensing of radiation reflected, scattered, or emitted back from the earth's surface through the atmosphere and what it tells us about the system being observed. These primarily top-down (climatological) and bottom-up (primarily physiological and population dynamics) data could help us interpret the changes in ecosystems. (Note that the terms top-down and bottom-up as used here are in a different context than used by ecologists.)

Satellite imagery has been used by Tucker et al. (1985) to develop greenness indices for North Africa based on spectral data. The goal was to estimate plant cover to predict regional favorability for locust development. Unfortunately, the indices do not tell us when host plants germinate, they are insensitive to a ground cover of less than 20%, and they do not give information on the species composition. In general, satellite predictions of rain events, their distribution, and quantity are imprecise, especially over large land masses such as North Africa. Another major problem is that data from satellites has been commercialized and is prohibitively expensive for many research groups. Hopefully, this will change as the technology of the cold war becomes more widely available.

Graphical Information Systems Technology

Graphical information systems (GIS) technology enables us to summarize information from many sources on a geographic basis, manipulate it, and create synthetic data that summarize the effects of several variables (e.g., growth indices) on small or large geographical scales. The applications of this technology is expanding rapidly in many disciplines worldwide, and its basis has been reviewed in an introductory text by Cracknell and Hayes (1991). A GIS system should have the capacity to provide interactive displays of cartographic or database information, provide different map projections and scales, facilitate the creation and manipulation of map symbols, provide easy development of digital map overlays, and allow for the conversion of raster (pixel or grid base) to vector and vice versa.

To apply GIS to the assessment of regional favorability for the development of a species in a time varying manner, we might first digitize a map of the geographic area with grid coordinates and store historical and real time data in a computer file by coordinates. By extrapolating between grid point, areas of equal or similar values may be identified and mapped. For example, we might have topographic data, soil type, nutrient status (e.g., N, P, K), salinity, drainage, rainfall, temperatures, and other factors associated with specific sites or areas. These data could be used to run models to update factors (e.g., soil water) and to generate growth indices of favorability for different species on a real time basis. For example, the study might be of land patterns (e.g., San Francisco Bay area), or of an analysis of wetlands to predict areas favorable for the breeding of various mosquitoes. On a larger scale, we might assess the whole of North Africa for the favorability of desert locust breeding (Fig. 11.4).

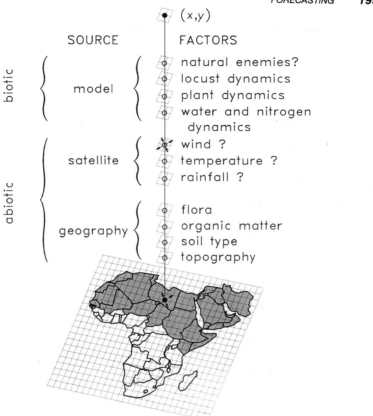

Figure 11.4. *Examples of site specific data required to run a plant–locust simulation in a GIS shell.*

FORECASTING

In North America, the pest phenology system (PETE) developed by Welch et al. (1978) to predict the proportion of the development that has occurred has been widely used. Holling et al. (1977) evaluated the regional dynamics of the spruce budworm in Nova Scotia. In this study, the population dynamics of the spruce budworm were computed for each grid point of a map, and their density at each time interval depicted as a point above the map surface. The points of the grid were connected to yield a time-varying dynamical three-dimensional picture of the status of the pest population across the region. This time series of maps is, by my definition, a geographic time-varying life table.

Characterizing Regional Favorability Using Physiological Indices

Fitzpatrick and Nix (1970) determined the areas of expected favorability for pastures during different seasons of the year based on average weather. Gutierrez et al. (1974b) modified the Fitzpatrick–Nix method and applied it to examine the effects of observed weather on the potential for development of the cowpea aphid (*Aphis craccivora* Koch) populations over a wide geographic and ecologically diverse area of South East Australia. Water availability is often limiting for pasture plants in the arid areas and frosts are limiting for the aphid. Weekly census data on winged aphid populations at 40 locations over two years (Fig. 11.5a) were used in the analysis. At the time, cowpea aphid had no effective natural enemies, and its populations were highly asynchronous across this vast climatically diverse area suggesting a mosaic of favorability of weather in time and space. On some occasions winged cowpea aphids were carried on the winds to locations where drought prevented plant growth, and conversely at other locations aphids were absent despite highly favorable conditions suggesting that no migration had occurred.

The average moisture indices $(\hat{MI})$ for the plant and the average temperature indices $(\hat{TI})$ for the aphid were computed using observed weather data from the various locations for periods when winged aphids were trapped for periods of three or more weeks (Fig. 11.5a). Quite surprisingly, the paired $(\hat{MI}, \hat{TI})$ values clustered not only for the target species, *A. craccivora*, but also for other species (e.g., *Rhopalosiphum padi* L. and *R. maidis* L., Fig. 11.5b) but at different places on the physiological phase space. The $\hat{MI}$ and $\hat{TI}$ are not independent, hence a bivariate normal statistical model was used to compute the 95% tolerance regions for each species (Fraser and Guttman 1956). In another analysis the data were weighted for the size of the population that developed at each site but this did not change the conclusions (Gutierrez and Yaninek 1983).

Aphis craccivora developed during spring and fall, *R. padi* flourished during cool wet periods, and *R. maidis* was favored by dry warm weather. Of course, simple observations could tell you this, and for a nearby area one could simply go look. However, if the area is large, one must surmise from far away at what time and place and how likely it would be for a species to reach outbreak levels. To do this, one could compute physiological indices to interpret recent weather from the species' point of view to determine whether further investigation is warranted. Aggregating the cowpea aphid phenology data based on physiological criteria (indices) demonstrates the utility of using second-order effects. The indices help determine where populations might develop, but not that they do develop or how large they will be. In the case of cowpea aphid or locusts, immigration to a favorable site may simply not occur.

Sutherst et al. (1991) linked generic "growth indices" in their development of a pest distribution GIS system (e.g., PESKY). The mathematics of their model has not been fully outlined, but their methods are derived from Fitzpatrick and Nix (1970). This computer program uses real time or historical weather to drive physiological indices that characterize the favorableness of areas for species. The system is purported to be very flexible and models are easily developed and included. Hughes and Maywald (1990) used an early version of the program CLIMEX to map the potential worldwide distribution of the Russian grain aphid [*Diuraphis noxia* (Mordvilko)], and the predictions accorded well with the known distribution of the pest in California.

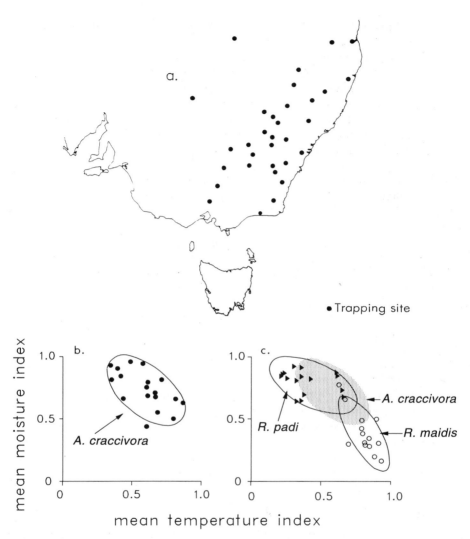

Figure 11.5. *Regional dynamics of the cowpea aphid (A. craccivora Koch) in South East Australia during 1968–1970: (a) distribution of Moericke yellow-pan water traps, (b) the average MI and TI indices for the periods when winged cowpea aphids were trapped at all locations, and (c) compared to two grass aphids [cf., Gutierrez et al. 1974b].*

The Desert Locust Problem

A formidable problem that requires such a regional analysis is the desert locust [*Schistocerca gregaria* (Forskal)]. Outbreaks of the desert locust and others in North Africa, the Middle East, and the Indian peninsula may cause severe crop losses and, in isolated areas, starvation. The desert locust is thought to be limited by weather with their densities remaining low for long periods until prolonged periods of favorability return allowing outbreaks to develop (Roffey and Popov 1968). An additional factor favoring the development of massive outbreaks is the occurrence of weather fronts that enhance migration success between favorable habitats. This allows the populations to gain growth momentum and to explode. This favorable continuum in time and space is rare and as a consequence outbreaks are the exception rather than the rule. The efficiency of regulatory biotic agents is not well documented for locust, but likely vertebrate predation on low-density populations slows but does not prevent outbreaks when favorable conditions return.

The last outbreak of desert locust occurred during the late 1980s, and $400 million was spent for chemical control (Symmons 1992). The efficacy of the effort is highly suspect (US Congress OTA 1990), and points to the need to predict the outbreak potential of locust across its vast range. This would help reduce crop losses by intervening early, reduce the effect to fragile ecosystems, such as oases, and possibly aid in implementing emergency relief work. Ultimately, real time regional dynamic models for the phenology of such pests must be developed to place the large scale pest control campaigns on a scientific basis. GIS work using ecological indices for this and other locusts was pioneered by M. Launois and colleagues in France.

Applications to Agriculture

In areas with good meteorological information, embedding realistic population models of crops, pest and natural enemies, and the dynamics of soil water and nutrients into a GIS system will enhance our ability to assess many of these interactions on a real time basis. Detailed models of alfalfa, cotton, bean, grape, rice, and cassava have been developed and extensively tested in California (see references cited therein), and their applications to a regional basis in a GIS system is in progress.

EPILOGUE

As with most early attempts to apply new technologies, large problems are apparent—the chasm between the technology that we can apply and our understanding of the underlying biology is often wide. The incautious may ignore this and allow the power of the computer to outstrip common sense and ignore Oster's (1981) admonition concerning fortuitous numerology, which was given in the introduction to this book. As the cold war thaws, improved satellites with greater resolution will come on line, greatly increasing our capacity to assess ecosystem dynamics. Hopefully, sound ecosystem management will be introduced on a global scale before the earth becomes a barren desert.

12

Ecosystem Sustainability

Advice given to President Franklin Pierce

This we know: the earth does not belong to man, man belongs to the earth. This we know: all things are connected like the blood which unites one family. All things are connected, what ever befalls the earth befalls the sons of the earth. Man did not weave the web of life; he is merely a strand in it. What ever he does to the web he does to himself.
—*Chief Seattle in 1854*

My grandfather's perspective

The country started going down in 1840 when the Anglos[1] started to arrive.
—*Julian Cordova*
Belen, New Mexico

A POINT OF VIEW

Ecologists seem not to agree on the meaning of the term *ecosystem sustainability*. This skeptical view is supported by the 1993 series of articles by several eminent ecologists [Costanza (1993); Ehrlich (1993); Fuentes, (1993); Hilborn and Ludwig (1993); Holling (1993); Ludwig (1993); Lee (1993); Mangel et al. (1993); Mooney and Sala (1993); Pitelka and Pitelka (1993); Policanski (1993); Rubenstein (1993); Salwasser (1993); Slobodkin (1993); Socolow (1993); Zedler (1993)] in their disparate responses to the thought provoking article by Ludwig et al. (1993) on the subject. The articles ran the gamut from optimism, where rational people can get together to solve ecosystem level problems, to pessimism, where we will not do it in time. The articles serve to illustrate how far apart ecologists are in defining and putting into practice this concept—*everyone knows what it is but each has a different definition.* Some

---

[1]Individuals may be of all races and color.

199

scientists (not in that group) privately express optimism that cheap power will enable us to solve all of our ecological problems and others are simply grateful to be old enough so that they will not be around when the real crunch comes. Among the less realistic optimists are some biotech colleagues who feel they can bioengineer their way out of agricultural and ecological problems (recall the green revolution), and that if we save the genes of endangered species we can bring them back later—shades of *Jurassic Park* (Crichton 1990). Even if possible, they seem not to have learned that much about ecosystems structure and function, or that much of higher animal behavior is learned and not innate.

I tend toward Ludwig's (1993) views on the subject of ecosystem sustainability because there are several mitigating factors when humankind is the shepherd of nature. As Ludwig urged, we must begin to talk about the litany of taboo subjects that impinge on ecosystem sustainability (*ecosystem survivorability*—my emphasis). Among the unlisted taboos that we need to talk about are our *ISMS*—those beliefs based on our economic, political, religious, social and personal perceptions of the world—our personal set of myths—our denials.

Not only do we struggle with ecosystem sustainability, but we also have to deal with the concomitant necessity of sustainable development (i.e., economic development), a term I think is an oxymoron, especially when we acknowledge that there are already too many people on earth, the population is still growing exponentially, and now the free enterprise of the global market is becoming the dominant economic paradigm replacing market places in developing areas that have human scale and dimension. The global market place can only increase the rate of ecosystem degradation as entrepreneurs seek to satisfy insatiable appetites for resources—a demand, if you will, for wealth accumulation. Consumers in capitalist societies are being trained to accept the unacceptable, and to have planning horizons that are barely more than day to day. Quarterly or yearly plans are the norm in economic houses who are at this moment buying and selling our children's future for a few "extra bucks." For global free-trade to flourish, we are using nonrenewable resources to move nonrenewable resources from one place to another. All too often resources are extracted against the best wishes of local inhabitants, and the practice continues until they are depleted or become too costly and can be replaced by a cheaper source from some place else where the cycle of exploitation is repeated. The notion of substitutability is pervasive, and as ecosystem degradation sets in, the ecological commons are increasingly someone else's responsibility—of eco-radicals (e.g., Carson 1962; Hardin 1968, van den Bosch 1978).

I recognize that my musings are outrageously polemical and politically incorrect, but time is of the essence for the niceties of prolonged discourse to prevail. I submit that the issue is not whether we can manage ecosystems in a sustainable manner, but rather how long we can keep the global subsystems going before they collapse like dominos due to our inevitable mismanagement in a cascading series of global catastrophies. Our record starting with the four basic Greek elements of air, earth, fire, and water, and now nature itself serves as ample modern reason for concern.

The theme of this book has been the interplay between the realized per capita supply of a resource and the per capita demand for it. This is the context of biology, but not of economics where the supply and demand of a resource may be all that is available. The economist Winters (1971) attempted to make analogies between the economies of nature and of humans. A species he equated to a firm, profit was analogous to the fitness of a species (i.e., what you invest in the next cycle), adaptedness to long-term firm survival, firm decision rules to the genetics of a species, and so on. I think he erred, as the comparison should have been species to species—of *Homo sapiens*, the ultimate predator, to any other predator in nature. Table 1 summarizes some analogous attributes.

What is absolutely clear is that only humans have escaped population regulation except during times of famine, war, and pestilence; only humans have insatiable demands for resources, and only they can devise ways to search for a resource until the very last unit has been exploited (e.g., dodo birds, passenger pigeons, sardines, tigers, whales, and yew trees). The genetic memory of natural selection has given all species except humans demands for resources that are in tune with their environment either intrinsically via natural selection or via control by natural enemies, antagonists, and competitors. It is in a climate of unrestrained demand that humans operate—demands that may be created for no other reason than the accumulation of wealth.

Quite frankly, like van den Bosch (1978), I think that humans are not smart enough to manage ecosystems; we are seriously flawed as a species: we are too greedy, too self-centered, too avaricious, and too absolute in our conviction

TABLE 1. Comparisons of the Economies of Species in Nature and Firms and Humans

Other Species	Modern Industrial Humans
• The species	• The firm or *Homo sapiens* sapiens
• Fitness	• Profit or *alternatively understanding*
• Adaptedness	• Long term fiscal stability *or a stable society*
• Mechanism—Genetic memory of natural selection	• Management rules *or social ethic*
• Goal—avoiding extinction	• Goal—wealth accumulation *or quality of life*
• Regulated by other species	• Escape regulation *or impose self regulation*
• Self-regulation—Buffeted by the environment	• Control our environment unsustainably *or live in sustainable environment*?
• Social structure adapted to meet challenges of the abiotic and biotic environment	• Social structure designed to increase wealth accumulation and dominate our environment (i.e., our *ISMS*) *or to enhance the common good*
• Demand for resources set by natural selection	• Demand potentially unconstrained by greed and technology *or set limits to wealth accumulation*
• Search limited to meet genetically programmed needs	• Search enhanced by modern technology— we can get the last one *or control our harvests at realistic sustainable levels*

that everything around us was placed there by God(s) to be exploited—and to some this includes each other.

Environmentalists are justly concerned about endangered species, but I think that the focus is wrong. I doubt that *nature* in the abstract cares whether this or that species goes extinct here or there. This includes humans. We should, however, be alarmed that the increasing numbers of species on the verge of extinction are indices of how badly the environment is being degraded by our activity. We should be concerned because if the current rate of global degradation continues, the species *Homo sapiens* may soon join the list of endangered species.

EPILOGUE

As the final bit of irony, it will be insects that polish the bones of the last of us to fall.

—**Robert van den Bosch, 1978**

Appendices: Some Tools for Applied Ecologists

Some Essential Mathematics for Biologists

by **Maurizio Severini**

Instuto di Fisica dell' Atmosfera
P.le Luigi Sturzo, 31
00144 Rome, Italy

Institute of Physics
Building 'E. Fermi'
Group of Meteorology
University of Rome 1
P.le A. Morro, 2
00185 Rome, Italy

FUNCTIONS

Many studies in biology have to do with relationships that exist between one collection of observations (objects). For example, one might observe that fecundity in females of a species increases with female size, or that developmental rate of a population is related to temperature. These relationships are represented in tables, graphs or equations and are called functions in the mathematical jargon. Through measurements and countings, observations must be transformed into numbers because mathematics and functions operate on these latter entities.

Given two collections of numbers X and Y, a function f is a correspondence which associates with each number of x of X, a number y of Y (Fig. A.I.1).

The numbers of X are called *independent variables*, those of Y *dependent variables*. The functional dependence that relates each number of X to one and only one number of Y is written:

$$y = f(x) \qquad \text{[A.I.1]}$$

An example of a function is the linear function, whose expression is:

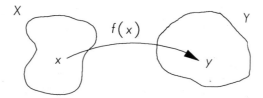

Fig. A.I.1.

$$f(x) = \alpha \cdot x + \beta \qquad [\text{A.I.2}]$$

where α and β are constant numbers. If the function f associates a number x of X and a number y of Y to one number z of a collection Z, it is called a function of two variables and it is written:

$$z = f(x, y) \qquad [\text{A.I.3}]$$

An example of this type of dependence is the Boyle's Law for an ideal gas. Let x be the number measuring the pressure of a gas, y its volume, and z its temperature. The law can be written:

$$z = \frac{1}{N \cdot R} \cdot x \cdot y \qquad [\text{A.I.4}]$$

where N is the number of moles of the gas and R is Avogadro's number.

There are a lot of other types of functions such as inverse functions, multivalued functions, complex functions, and so on, but their introduction goes beyond the limits of these preliminaries.

CONTINUITY

The notion of continuity deals with the property of connectedness of a function. If a function is represented by a graph, then its continuity can be simply determined by noting that the graph has no jumps or holes. Consider the following examples (Fig. A.I.2). There are two functions $f_1(x)$ and $f_2(x)$. The first function is continuous in x_0 whereas the second one is not continuous in the same point; it is said to be discontinuous at x_0. What is the difference? In both cases the functions are defined (each has a finite value) at x_0, but this is not sufficient for continuity. Continuity states that the function f_1 approaches the value $f(x_0)$, from both left and right of x_0. In contrast $f_2(x)$ shows this behavior and x_0 only from the left side. In the more rigorous symbolism of mathematics, the concept of *limit* is used to define continuity as follows:

$$\lim_{x \to x_0} f(x) = f(x_0) \qquad [\text{A.I.5}]$$

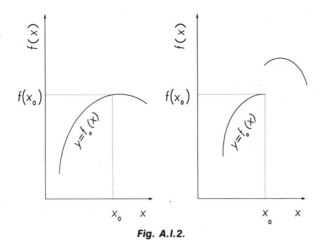

Fig. A.I.2.

An important consequence of the previous definition is that if $f(x)$ and $g(x)$ are two functions that are continuous at x_0, their sum, difference, and product are still continuous at x_0. Their ratio, $f(x)/g(x)$, is also continuous in x_0 provided $g(x_0) \neq 0$.

DIFFERENTIATION

Differentiability of a function is concerned with its smoothness. For instance, given a function $f(x)$ continuous at x_0, let us consider an *increment f* of $f(x)$ defined as:

$$\Delta f = f(x) - f(x_0) \qquad \text{[A.I.6]}$$

corresponding to an increment Δx of x around x_0

$$\Delta x = x - x_0. \qquad \text{[A.I.7]}$$

The ratio of these two increments,

$$\frac{\Delta f}{\Delta x} \qquad \text{[A.I.8]}$$

which is called an *incremental ratio* giving the proportional variation of the dependent variable with respect to the independent one. The limit of this ratio, if it exists as the independent variable x tends to the value x_0, defines the *derivative* of the function $f(x)$ with respect to x at the point x_0.

$$\lim_{x \to x_0} \frac{f(x) - f(x_0)}{x - x_0} = f'(x_0) \qquad\qquad [\text{A.I.9}]$$

Because of its paramount importance in mathematics, the derivative has a certain number of equivalent expressions; here are some of them using the different notions you commonly encounter:

$$[Df(x)]_{x = x_0} = \left.\frac{df}{dx}\right|_{x = x_0} = \lim_{x \to x_0} \frac{\Delta f}{\Delta x} \qquad\qquad [\text{A.I.9a}]$$

The result $f'(x_0)$ is a number, but if x_0 is considered to vary in x and if its derivative exists for all x, a new function $f'(x)$ is obtained through the operation indicated by (9), which is called the *derivative function* of $f(x)$:

$$\frac{df(x)}{dx} = f'(x) \qquad\qquad [\text{A.I.10}]$$

Differentials dx and df are introduced from increment's definitions [A.I.6] and [A.I.7]

$$\lim_{x \to x_0} \Delta x = dx \qquad\qquad [\text{A.I.11}]$$

and

$$\lim_{x \to x_0} \Delta f = df \qquad\qquad [\text{A.I.12}]$$

Often in computational problems, derivatives [A.I.9a] are considered as ratios of differentials, even if this procedure is not strictly correct from the theoretical point of view.

A first consequence of defintion [A.I.9] and [A.I.10] are the basic rules of differentiation. We use the shorter notion $\{$e.g., $D[f(x)]\}$ rather than df/dx for convenience.

$$D[f(x) + g(x)] = D[f(x)] + D[g(x)]$$

$$D[\alpha \cdot f(x)] = \alpha \cdot D[f(x)]$$

$$D[f(x) \cdot g(x)] = D[f(x)] \cdot g(x) + f(x) \cdot D[g(x)]$$

$$D\left[\frac{f(x)}{g(x)}\right] = \frac{D[f(x)] \cdot g(x) - f(x) \cdot D[g(x)]}{[g(x)]^2} \qquad\qquad [\text{A.I.13}]$$

the last expression is not valid for that value of x for which $g(x) = 0$. A special rule is the chain-rule for evaluating nested functions.

$$D[f(g(x))] = D[f(g(x))]D[g(x)]$$

A second consequence of [A.I.9] and [A.I.10] is the ability to calculate the derivatives of the main mathematical functions that are also important in biology and ecology.

$$D[\alpha] = 0$$

$$D[x] = 1$$

$$D[\alpha \cdot x + B] = \alpha$$

$$D[x^\alpha] = \alpha \cdot x^{\alpha - 1} \text{ if } \alpha \neq 0$$

$$D[\alpha^x] = \alpha^x \ln \alpha \text{ if } \alpha \neq 0$$

$$D[\log_\alpha x] = \frac{1}{x} \log_\alpha e$$

$$D[e^{\alpha x}] = e^{\alpha x} \text{ if } \alpha \neq 0$$

$$D[\ln x] = \frac{1}{x}$$

$$D[\sin x] = \cos x$$

$$D[\cos x] = -\sin x \qquad \text{[A.I.14]}$$

where $e = 2.71828\ldots$ (not rational) is the Napier's number, and log and ln indicate logarithms to the base 10 (Briggs) and base e (natural), respectively.

A third consequence of [A.I.9] and [A.I.10] is the following. One can consider that the "new" function $f'(x)$ is "generated from" *the "old one"* (i.e., $f(x)$). From this point of view the function $f(x)$ is called the *primitive function* of $f'(x)$ and expression [A.I.10] is considered not merely a definition of the derivative, but also as an equation connecting $f'(x)$ with its primitive $f(x)$. From this point of view, expression [A.I.10] is called a *differential equation*.

INTEGRATION

Integration is roughly the reverse operation of differentiating a function. This operation is important because there are many theoretical and practical problems both in physics and in ecology where the derivative function is known and the primitive one is the unknown, which must be calculated. [In these cases the derivative function is indicated by $f(x)$ instead of $f'(x)$ to remember that it is a given function, and the primitive function is indicated by the new symbol $F(x)$.] The differential equation [A.I.10] shows the more usual aspect [A.I.15]:

$$\frac{dF(x)}{dx} = f(x) \qquad \text{[A.I.15]}$$

This equation has to be solved with respect to the unknown function $F(x)$ and the operation leading to this result is called *integration* and is indicated as:

$$F(x) = \int f(x)\, dx + C \qquad [A.I.16]$$

where C is an arbitrary constant of integration. The presence of this arbitrary constant justifies the name of *indefinite integral* given to the new symbol introduced in [A.I.16]. Integration is the search for a function $F(x)$, which is the primitive of a given $f(x)$, while $F(x)$ is by definition [A.I.15] the function whose derivatives is $f(x)$.

From equation [A.I.15] and the definition of indefinite integration [A.I.16], it is possible to establish two elementary properties of the integrals involving the sum of function and the product of a function with a scalar.

$$\int [f(x) + g(x)]\, dx = \int f(x)\, dx + \int g(x)\, dx$$

$$\int \alpha \cdot [f(x)]\, dx = \alpha \int f(x)\, dx \qquad [A.I.17]$$

It is important to note that multiplications and divisions with integrals do not have general rules.

The integration of a function may not be an easy operation, but in a number of cases it is possible to take advantage of the calculus of derivatives [A.I.14]. Integrals drawn up in this manner are called *fundamental integrals*, and some of them are listed here.

$$\int \alpha \cdot dx = \alpha \cdot x + C$$

$$\int x^{\alpha}\, dx = \frac{x^{\alpha+1}}{\alpha+1} + C \qquad (\alpha \neq -1)$$

$$\int \frac{1}{x}\, dx = \ln x + C$$

$$\int \alpha^{x}\, dx = \frac{\alpha^{s}}{\ln \alpha} + C$$

$$\int e^{x}\, dx = e^{x} + C$$

$$\int \sin x\, dx = -\cos x + C$$

$$\int \cos x\, dx = \sin x + C \qquad [A.I.18]$$

The arbitrary constant of integration C shows that the result of an indefinite integration is not univocal; it must be determined by some *auxiliary condition* for the primitive function, such as an *initial condition*.

The operation of *definite integration* of the function $f(x)$ between two values of x, say $x = a$ and $x = b$, can be introduced from the indefinite one as

$$I_{ab} = \int_a^b f(x) \, dx = F(b) - F(a). \qquad \text{[A.I.19]}$$

The new integral symbol together with the two extremes a and b is called *definite integral*. The value of the definite integral [A.I.19] I_{ab} is a number, that is, the difference between the values of primitive function $F(x)$ at b and at a, respectively.

If the symbol of the independent variable x is substituted in $f(x)$ by a new one γ, the new expression $f(\gamma)$ is obtained. The definite integral of this "new" function between $\gamma = a$ and $\gamma = b$ with respect to γ will be the same number as [A.I.19], that is, I_{ab}:

$$I_{ab} = \int_a^b f(\gamma) \, d\gamma = F(b) - F(a) \qquad \text{[A.I.19a]}$$

Now let us consider the upper extreme of integration b varies in the collection of numbers x and X such that $b = x$, then the value I_{ax} of the definite integral [A.I.19a] will become a function of the upper limit x:

$$I_a(x) = \int_a^x f(\gamma) \, d\gamma = F(x) - F(a) \qquad \text{[A.I.20]}$$

Operation [A.I.20] can be considered as a *definite integration with respect to the upper limit*. Consideration of expression [A.I.16] will show that the last operation gives the same result as the indefinite integration $F(x)$, that is, the primitive function of $f(x)$.

$$F(a) = C$$

$$C + I_a(x) = F(x)$$

PARTIAL DIFFERENTIATION AND DOUBLE INTEGRATION

Partial differentiation and double integration arise from the application of differentiation and integration operations to multivariables functions [A.I.3].

A function $f(x, y)$ of two independent variables x and y can be represented as a surface in a three dimensional space $O(x, y, z)$ {that is, *the collection of point* $[x, y, f(x, y)]\}$. The couple (x, y) identifies a point $P = (x, y)$ in the coordinate plane $z = 0$, and the function $f(x, y)$ can be defined as a function of this point, that is, $f(P)$. The value $z = f(P)$ moves on a surface in the space when P moves on the $z = 0$ plane; in this way a surface is defined that is the geometric representation of $f(x, y)$.

Often it is useful to consider such functions as two functions of one variable. This results when one of the two independent variables is taken as a constant. For example, if the second variable y is assumed to be a constant y_0, $f(x, y)$ becomes a one variable function:

$$f_{y_0}(x) = f(x, y_0) \qquad [A.I.21a]$$

If the variable x is assumed to be a constant x, so that $f(x, y)$ becomes the second one variable function:

$$f_{x_0}(y) = f(x_0, y) \qquad [A.I.21b]$$

The condition of continuity of a two variable function $f(x, y)$ at a point $P_0 = (x_0, y_0)$ comes from that of a single variable one. To assess the continuity of the surface $f(x, y)$ in its point of coordinate $z_0 = f(x_0, y_0)$, it is necessary that when the point $P = (x, y)$ approaches the point $P_0 = (x_0, y_0)$ from all directions on the plane (x, y), the point $z = f(x, y)$ must approach z_0 on the surface without "jumps" or "holes" (Fig. A.I.3). This behavior is expressed in mathematical symbolism as:

$$\lim_{(x,y) \to (x_0,y_0)} f(x, y) = f(x_0, y_0) \qquad [A.I.22]$$

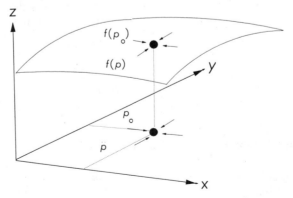

Fig. A.I.3.

For a two variable continuous function, it is possible to define two different partial increments: $\Delta_x f$ and $\Delta_y f$, one with respect to x and one to y

$$\Delta_x f = f(x, y_0) - f(x_0, y_0) \qquad \text{[A.I.23a]}$$

$$\Delta_y f = f(x_0, y) - f(x_0, y_0) \qquad \text{[A.I.23b]}$$

according to the increments of the two independent variables:

$$\Delta x = x - x_0 \qquad \text{[A.I.24a]}$$

$$\Delta y = y - y_0 \qquad \text{[A.I.24b]}$$

As for the single-value functions, the limit of the incremental ratio when it exists is called the derivative. The difference now is that there are *two partial incremental ratios* and thus two *partial derivatives* defined as:

$$[D_x f(x, y)]_{x = x_0, y = y_0} = \lim_{x \to x_0} \frac{f(x, y_0) - f(x_0, y_0)}{x - x_0} \qquad \text{[A.I.25a]}$$

$$[D_y f(x, y)]_{x = x_0, y = y_0} = \lim_{y \to y_0} \frac{f(x_0, y) - f(x_0, y_0)}{y - y_0} \qquad \text{[A.I.25b]}$$

There are a number of equivalent symbolic expressions for the partial derivatives

$$[D_x f(x, y)]_{x = x_0, y = y_0} = \left[\frac{\partial f}{\partial x} \right]_{x = x_0, y = y_0} = f_x(x_0, y_0) \qquad \text{[A.I.26a]}$$

$$[D_y f(x, y)]_{x = x_0, y = y_0} = \left[\frac{\partial f}{\partial y} \right]_{x = x_0, y = y_0} = f_y(x_0, y_0) \qquad \text{[A.I.26b]}$$

taken from the definition of *partial differentials*

$$\lim_{x \to x_0} \Delta x = \partial x \qquad \text{[A.I.27a]}$$

$$\lim_{y \to y_0} \Delta y = \partial y \qquad \text{[A.I.27b]}$$

$$\lim_{x \to x_0} \Delta_x f = \partial f \qquad \text{[A.I.28a]}$$

$$\lim_{y \to y_0} \Delta_y f = \partial f \qquad \text{[A.I.28b]}$$

The partial differentials refer to one of the function's independent variable;

and therefore are indicated by the symbol ∂, which is different from d introduced earlier for ordinary differentials.

By allowing x_0 and y_0 to vary, the *partial derivative functions* are finally obtained from [A.I.25a], [A.I.25b], [A.I.26a], and [A.I.26b]

$$\frac{\partial f(x, y)}{\partial x} = f_x(x, y) \qquad \text{[A.I.29a]}$$

$$\frac{\partial f(x, y)}{\partial y} = f_y(x, y) \qquad \text{[A.I.29b]}$$

Expressions [A.I.29a] and [A.I.29b] can be viewed as *partial differential equations* that arise when the functions $f_x(x, y)$ and $f_y(x, y)$ are known and the two variables function $f(x, y)$ is unknown. In this case we are reintroduced, in a very natural way, to the problem similar to that of finding the primitive function of a one variable function.

Let us consider just one of the two last equations. If $f_x(x, y)$ is known, let us indicate it by $f(x, y)$. Functions with two variables have two possible partial derivatives, and consequently there may exist *two functions* $U(x, y)$ and $V(x, y)$ whose partial derivatives (e.g., the first with respect to x and the second to y) are identical to $f(x, y)$:

$$\frac{\partial U(x, y)}{\partial x} = f(x, y) \qquad \text{[A.I.30a]}$$

$$\frac{\partial V(x, y)}{\partial y} = f(x, y) \qquad \text{[A.I.30b]}$$

The functions $U(x, y)$ and $V(x, y)$ are called *partial primitive functions* of $f(x, y)$ or, more precisely, the first with respect to x and the second with respect to y.

Partial integration is the operation that enables us to find a partial primitive function using *indefinite partial integration*:

$$U(x, y) = \int f(x, y) \, dx + C_1 = \int_a^x f(\gamma, y) \, d\gamma \qquad \text{[A.I.31a]}$$

$$V(x, y) = \int f(x, y) \, dy + C_2 = \int_b^y f(x, \gamma) \, d\gamma \qquad \text{[A.I.31b]}$$

The *indefinite partial integration* is made by considering one variable as a constant and integrating with respect to the other one. The *definite partial integration with respect to the upper limit* is calculated by substituting the variable of integration with a *dummy variable* γ, and then integrating with respect to the former with the remaining variable considered a constant.

Finally, two subsequent partial integrations of a two variable function leads to the definition of *double integration* and to that of *double integrals*, as follows:

$$I(x, y) = \int_a^x \left[\int_b^y f(\gamma, \varepsilon) \, d\varepsilon \right] d\gamma = \int_a^x \int_b^y f(\gamma, \varepsilon) \, d\varepsilon \, d\gamma \quad \text{[A.I.32]}$$

Two dummy variables are introduced to show that the double integration of $f(x, y)$ is simply a sequence of two partial integrations. The result of the first definite partial integration with respect to one variable is successively partially integrated with respect to the other one.

DIFFERENTIAL EQUATIONS

Equation [A.I.11] is a particular case and perhaps the easiest differential equation whose solution can be found by elementary integration.

Ordinary differential equations in their broadest definition are equations involving derivatives of an unknown function of a single variable. Here are some examples of differential equations.

$$\frac{df(x)}{dx} + x \cdot [f(x)]^2 = 2x$$

$$\frac{df(x)}{dx} = \alpha \cdot f(x)$$

$$\frac{df(x)}{dx} = \alpha \cdot f(x) \cdot [\beta - f(x)]$$

$$\frac{df(x)}{dx} = \alpha \cdot f(x) + \beta \cdot g(x) + \gamma \cdot h(x) \quad \text{[A.I.33]}$$

The last three equations are well known in population ecology. By substituting time (t) for the independent variable x in the last three equations, we see that they are the exponential growth equation (i.e., the second), the logistic equation (the third), and the enzyme kinetic equation (the fourth), respectively.

Partial differential equations are equations containing partial derivatives of two or more variables of unknown function. Here are two examples:

$$\frac{\partial f(x, y)}{\partial x} = g(x) \cdot f(x, y) + h(y)$$

$$h(y) \frac{\partial f(x, y)}{\partial x} - g(x) \frac{\partial f(x, y)}{\partial y} = 0 \quad \text{[A.I.34]}$$

Differential equations are very important in physics and in biology because in the study of a large number of phenomena it is easier to get experimental information and relationships about the derivative(s) of an unknown function than of the function itself. In these cases, it is always necessary to write and solve one or more differential equations, that is, to find $f(x)$ or $f(x, y)$ starting from some relationship(s) connecting the derivative(s) of functions such as like examples [A.I.33] and [A.I.34]. It is also worth noting that the main difference between an algebraic and a differential equation is that a solution of the first is a number, whereas a solution of the second is a function.

MATRICES AND VECTORS

A matrix of real numbers is a table like the following written in three equivalent forms:

$$\mathbf{A} = (\alpha_{j,k}) = \begin{pmatrix} a_{1,1} & a_{1,2} & \cdots & a_{1,K} \\ a_{2,1} & a_{2,2} & \cdots & a_{2,K} \\ \cdot & \cdot & \cdot & \cdot \\ a_{J,1} & a_{J,2} & \cdots & a_{J,K} \end{pmatrix} = \mathbf{A}(J \times K) \quad \text{[A.I.35]}$$

where all the elements

$$\alpha_{j,k} \qquad \text{[A.I.36]}$$

with $j = 1, 2, 3, \ldots, J$

$\quad k = 1, 2, 3, \ldots, K$

are real numbers. Matrix $\mathbf{A}$ is called an $(J \times K)$ matrix with entries $\alpha_{j,k}$.

If $J = 1$ matrix $\mathbf{A}$ becomes a row vector of K components:

$$(\alpha_{j,k}) = \upsilon = (\alpha_{1,1}, \alpha_{1,2}, \ldots, \alpha_{1,K}) \qquad \text{[A.I.37]}$$

and if $K = 1$ the same matrix becomes a column vector of J components:

$$(\alpha_{j,k}) = \omega = \begin{pmatrix} \alpha_{1,1} \\ \alpha_{2,1} \\ \cdot \\ \alpha_{J,1} \end{pmatrix} \qquad (38)$$

when $J = K$, then matrix $\mathbf{A}$ is called a *square matrix* of K rows and K columns.

Consider a second matrix **B** defined as:

$$B = b_{n,m} \qquad \text{[A.I.39]}$$

with:

$$m = 1, 2, 3, \ldots , M$$
$$n = 1, 2, 3, \ldots , N$$

If $M = J$ and $N = K$ the *sum of two matrices* **A** and **B** can be defined as:

$$A + B = (a_{j,k} + b_{j,k}) \qquad \text{[A.I.40]}$$

where the following condition is applied: $m = j$ and $n = k$. This means that the first element of matrix **A** must be summed to the first element of matrix **B**, the second element of the first row of **A** must be summed to the element in the same position in **B**, and so on.

If is α is a number, the product of a number times a matrix is defined as:

$$\alpha \cdot A = (\alpha \cdot a_{j,k}) \qquad \text{[A.I.41]}$$

i.e., the new matrix has the same elements of the old one multiplied by the number α.

If $K = M$ the product of two matrices (rows by columns) can be defined as:

$$A \cdot B = \left(\sum_{k=1}^{K} a_{j,k} \cdot b_{k,n} \right) = C = (c_{j,n}) \qquad \text{[A.I.42]}$$

The symbol Σ means sum and is widely used in mathematics, for example:

$$\sum_{k=1}^{4} a_k = a_1 + a_2 + a_3 + a_4$$

In expression [A.I.42] to obtain $c_{1,1}$, the first element of the first row of **A** is multiplied by the first element of the first column of **B** and the result is summed with the product between the second element of the first row and the second element of the second column, and so on, to the last elements of the first row and of the first column. Then this operation is repeated for all the rows of **A** and the columns of **B**. To make clearer the operation and the application of the symbol Σ, the explicit components of some elements of matrix **C** are given.

$$c_{1,1} = \sum_{k=1}^{K} a_{1,k} \cdot b_{k,1} = a_{1,1} \cdot b_{1,1} + a_{1,2} \cdot b_{2,1} + \cdots + a_{1,k} \cdot b_{k,1}$$

$$c_{1,2} = \sum_{k=1}^{K} a_{1,k} \cdot b_{k,2} = a_{1,1} \cdot b_{1,2} + a_{1,2} \cdot b_{2,2} + \cdots + a_{1,k} \cdot b_{k,2}$$

$$c_{2,1} = \sum_{k=1}^{K} a_{2,k} \cdot b_{k,1} = a_{2,1} \cdot b_{1,1} + a_{2,2} \cdot b_{2,1} + \cdots + a_{2,k} \cdot b_{k,1}$$

$$c_{J,N} = \sum_{k=1}^{K} a_{J,K} \cdot b_{k,N} = a_{J,1} \cdot b_{1,N} + a_{J,2} \cdot b_{2,N} + \cdots + a_{J,k} \cdot b_{k,N}$$

The rule for matrix multiplication is that multiplication can occur if $K = M$

$$\mathbf{A}(J \times K) \cdot \mathbf{B}(M \times N) = \mathbf{C}(J \times N),$$

but not if $K \neq M$.

When $N = 1$, matrix $\mathbf{B}$ defined in [A.I.39] reduces to a column vector according to [A.I.38].

$$b_{m,1} = \omega = \begin{pmatrix} b_{1,1} \\ b_{2,1} \\ \cdot \\ b_{M,1} \end{pmatrix}$$

The product $\mathbf{A} \cdot \mathbf{B}$ now becomes a product of a matrix by a column vector:

$$\mathbf{A} \cdot \omega = \left(\sum_{k=1}^{K} a_{j,k} \cdot b_{k,1} \right) = (c_{j,1}) = \omega_1 \qquad \text{[A.I.43]}$$

whose result is a new column vector ω_1 of J components. If $\mathbf{A}$ is a square matrix ($J = K$) expression [A.I.43] gives the result:

$$\mathbf{A} \cdot \omega = \omega_1$$

The product of a $(K \times K)$ square matrix by a vector $(K \times 1)$ gives a new vector with the same number of components K as the old one. This operation is often interpreted as a transformation of the old vector ω into a new one ω_1 performed by the square matrix $\mathbf{A}$, and for this reason it is called a *transformation matrix*.

A Mathematical Overview of Population Models[1]

It is not mathematics that limits us, it is our ignorance of events in the field.
—N. E. Gilbert

INTRODUCTION

Biology is still largely a descriptive science, but mathematics has become an increasingly useful tool for investigating complex interactions. However, unlike physics and chemistry, biological units possess considerable variability and often behavior that has made the development of theoretical biology difficult. The formulation of reasonably realistic laws describing the behavior of all individuals of a particular species is likely an impossible task.

The fundamental methodology that needs to be employed in biology is the formulation of mathematical models, the comparison of their predictions with field and/or laboratory data, and, at a qualitative level, to ecological theory. However, when applying mathematics in biology there is a real danger of ignoring important biological details simply for the sake of applying refined theory and applying powerful computational tools to a simplified tractable mathematical description of biological processes. The results obtained may be misleading and unrealistic.

In this chapter, we review some of the important mathematical concepts that have been used in population modeling, and especially to models of field systems. In previous chapters, we stressed the inclusion of physiological processes at the per capita level in the model to increase realism.

---

[1]Written in cooperation with H. Y. Wang, formerly of my research group.

POPULATION MODELS

Single Species

Total Population Models. A population model attempts to predict the number of organisms at any time t. In very simple models, all individuals in a population are considered identical. This limitation is, of course, not realistic, but because these models form a basis for more of the development of complicated and useful models, it is essential to discuss them first (see also Pielou 1969).

Discrete Form—Difference Equation. Suppose we are conducting an experiment, and at each increment of time, Δt, the number of organisms is counted. Let $N(t)$ = the number of organisms at time t. If no births or deaths occur, then the number at $t + \Delta t$ is

$$N(t + \Delta t) = N(t)$$

Suppose the number of births and deaths are proportional to the population size. Let λ and μ be the birth and death rates, respectively, then

$$N(t + \Delta t) = N(t) + \lambda \cdot \Delta t \cdot N(t) - \mu \cdot \Delta t \cdot N(t)$$
$$= (1 + \lambda \cdot \Delta t - \mu \cdot \Delta t)N(t) \qquad [\text{A.II.1}]$$

where Δt is chosen to be small so that births and deaths occur simultaneously. If the number of organisms [$N(t_o)$] at time $t = t_0$ is given, subsequent population numbers at time $t = t_0 + k\Delta t$, where k is any positive integer and can be computed. The number $N(t_0)$ is called the initial condition.

Continuous Form—Differential Equation. By rearranging [A.II.1], we obtain

$$\frac{N(t + \Delta t) - N(t)}{\Delta t} = (\lambda - \mu)N(t) \qquad [\text{A.II.2}]$$

Take the limit as Δt goes to 0, we have the differential equation

$$\frac{dN(t)}{dt} = (\lambda - \mu)N(t) \qquad [\text{A.II.3}]$$

The solution of [A.II.3] is

$$N(t) = N(t_0)e^{(\lambda - \mu)t}$$

hence given initial conditions $N(t_0)$ at time t_0, $N(t)$ can be computed for all $t > t_0$.

Case 1: $\lambda - \mu > 0$ or more births than deaths

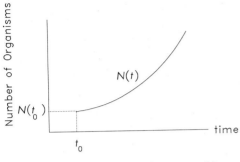

Population numbers increase without bounds

Case 2: $\lambda - \mu < 0$ or more deaths than births

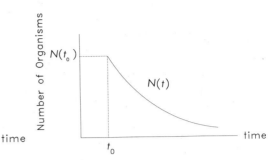

Population numbers decrease to zero

In the previous sections, we have considered only simple birth and death processes. The probability that an organism will reproduce or die remains constant and is independent of the population size. In reality, most species operate in a restricted environment and must eventually be limited by a shortage of resources, which could be food, shelter, mates, or other essential requisites. In such environments, birth and death rates are not constants, and must at a minimum be a function of the population number, $N(t)$. Let us combine the birth and death functions into one, say $\lambda(t)$. Then [A.II.3] becomes

$$\frac{dN(t)}{dt} = \lambda(N(t))N(t) \qquad \text{[A.II.4]}$$

The simplest form that $\lambda(N)$ can take is linear (Lotka 1925), that is,

$$\lambda(N) = a - bN \qquad a, b > 0 \qquad \text{[A.II.5]}$$

This is a general form of the model previously studied, [A.II.3], but here $a = (\lambda - \mu)$, $b = 0$. Equation [A.II.4] is the well-known logistic model

$$\frac{dN(t)}{dt} = (a - bN)N = aN(t) - bN(t)^2 \qquad \text{[A.II.6]}$$

with solution

$$N(t) = \frac{a}{b}\left[1 + e^{-a(t - t_o)}\frac{a/b - N(t_o)}{N(t_o)}\right]^{-1} \qquad \text{[A.II.7]}$$

Because $e^{-at} \to 0$ as $t \to \infty$, we get $\lim_{t \to \infty} N(t) = a/b$ (i.e., the environmental carrying capacity). Graphically, [A.II.5] and its solution are depicted below.

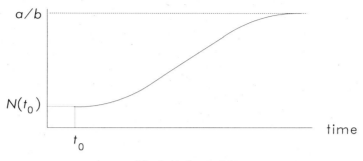

The logistic model

Additional complications may be introduced into [A.II.5] and other forms may be used. In general, such models are not realistic because all of the biology is subsumed in two constants.

Age Specific Models. The above population models have assumed that all organisms of a species are physiologically identical regardless of size or age. Individuals in different age brackets, however, may behave quite differently. Thus, conceptually, mathematical models would be more realistic if they incorporate age dependence. This leads to the well-known Leslie (1945, 1948) and McKendrick (1926) and von Foerster (1959) models.

Discrete Form: Leslie's Matrix. At each time t, the population of the organism is divided into n groups according to age, say, $N_{1,t}, N_{2,t}, \ldots, N_{n,t}$. Denote the difference of age between two successive groups by Δa. The birth and death rates can be age dependent. Let λ_i and μ_i be the birth and death rates for age group $i\Delta t$. If we choose Δt to be equal to Δa, then we obtain the following survivorship relationship:

$$N_{i+1,t+\Delta t} = (1 - \mu \cdot \Delta t)N_{i,t} \quad \text{for} \quad i = 1, 2, \ldots, n-1 \quad \text{[A.II.8]}$$

and population birth rate $N_{1,t+\Delta t}$

$$N_{1,t+\Delta t} = \sum_{i=1}^{n} \lambda_i \Delta t N_{i,t}$$

In equivalent matrix notation, we have [A.II.9]

$$
\begin{pmatrix} N_{1,t+\Delta t} \\ N_{2,t+\Delta t} \\ \vdots \\ N_{n,t+\Delta t} \end{pmatrix}
$$

$$
= \begin{pmatrix} \lambda_1\Delta t & \lambda_2\Delta t & \cdots & & \lambda_n\Delta t \\ 1-\mu_1\Delta t & 0 & & & 0 \\ 0 & 1-\mu_2\Delta t & \cdot & & \\ \vdots & & \cdots & & \vdots \\ 0 & & & 1-\mu_{n-1}\Delta t & 0 \end{pmatrix} \begin{pmatrix} N_{1,t} \\ N_{2,t} \\ \vdots \\ N_{n,t} \end{pmatrix}
$$

$$[\text{A.II.9}]$$

For all practical purposes, the last age bracket is the maximum age that individuals in the population are assumed to live. Given an initial population density vector

$$
\begin{pmatrix} N_{1,t_0} \\ N_{2,t_0} \\ \vdots \\ N_{n,t_0} \end{pmatrix}
$$

from [A.II.8] or [A.II.9], subsequent population density vectors can be solved at time $t = t_0 + k\Delta t$ where k is any positive integer. To arrive at [A.II.9], we have assumed that $\Delta t = \Delta a$ so that the notation is simplified. In practice, it is possible to incorporate varying Δt and Δa at the expense of complicating the mathematics.

Continuous Form: McKendrick–von Foerster Equation. The continuous counterpart of Leslie's discrete model is the McKendrick–von Foerster model:

$$
\frac{\partial N}{\partial t} + \frac{\partial N}{\partial a}\frac{da}{dt} = -\mu(\cdot)N(t, a) \qquad [\text{A.II.10}]
$$

where $N(t, a)$ is the population density function depending on time t and age a, $\mu(\cdot)$ is the death function depending on various parameters dictated by the problem and $da/dt = 1$. Equation [A.II.8] needs two initial conditions, $N(0, a)$ and $N(t, 0)$, to guarantee the uniqueness of its solution. $N(0, a)$ is the initial age distribution of the population. $N(t, 0)$ is the density of the newborn.

To illustrate the fact that [A.II.10] is the continuous version of Leslie's model, we discretize it. We approximate *dN/dt* and *dN/da* by finite differences

$$\frac{N(t + \Delta t, a + \Delta a) - N(t, a + \Delta a)}{\Delta t} \quad \text{and} \quad \frac{N(t, a + \Delta a) - N(t, a)}{\Delta a}$$

respectively. If we take Δt to be equal to Δa and substitute into [A.II.10], it becomes

$$N(t + \Delta t, a + \Delta a) = N(t, a) - \mu(\cdot)\Delta t N(t, a)$$

which is equivalent to the Leslie model. The point dependence of the model is illustrated below.

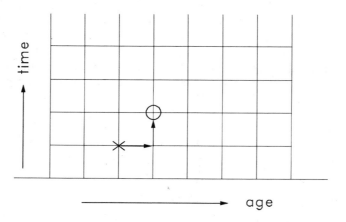

The McKendrick–von Foerster's equation can either be solved by *numerical techniques* mentioned above, or it can be solved analytically as follows: Equation [A.II.10] is equivalent to two ordinary differential equations.

$$\frac{da}{dt} = 1 \qquad\qquad \text{[A.II.10a]}$$

$$\frac{dN}{da} = -\mu(\cdot)N(t, a) \qquad\qquad \text{[A.II.10b]}$$

We assume here that the death rate depends only on t and a, that is, $\mu(\cdot) = \mu(t, a)$. The solution of [A.II.10a] is

$$a = t - t_o$$

where t_0 is the time of initiation. By separation of variables, [A.II.10b] is solved

$$\ln N = -\int_0^a \mu(x + t_0, x)\, dx + \ln N(t_0, 0)$$

Taking the antilog yields

$$N(t, a) = N(t_0, 0) \exp\left[-\int_0^a \mu(x + t_0, x\, dx\right]$$

$$= N(t - a, 0) \exp\left[-\int_0^a \mu(t - a + x, x)\, dx\right]$$

Continuous Form: Distributed Delay Models. In the above models, Δt has been assumed equal to Δa, but in reality the developmental times of individuals initiated at the same time may have a characteristic mean and variance. To model such processes requires a different model such as that proposed by Manetsch (1976) and Vansickle (1977). This model simulates the dynamics of a population wherein members of a cohort initiated at time t_0 have developmental times through a life stage characterized by one of a family of Erlang distributions. Severini et al. (1990) outline the theoretical basis of this model and many of his arguements are outlined in the Appendix of Chapter 9. Curry and Feldman (1987) define the problem in general terms. Of course, other distributions (binomial, etc.) could just as easily be used and Plant and Wilson (1986) review this problem, but only the Erlang distribution has a sound theoretical basis for use in such delay processes (Severini et al. 1990). Here we focus on the widely used Erlang distribution because of its theoretical basis.

In this model, as with the Leslie and McKendrick–von Foerster models, the different life stages may be modeled separately and the index for age can be used to delimit the life stages within a vector $(N_1, N_2, \ldots, N_k)$. The model is deterministic and only simulates stochastic development. Assume we are modeling the number dynamics of a population in the absence of mortality with cohorts N_i of age $i = 1, 2, \ldots, k$ and of age width D/k, where D is the mean developmental time in physiological time units. Here, we are assuming a time invariant form of the model, though a time-variant form may also be used. In the first and more widely used approach, D can be considered as a thermal constant, that is, the transit time through a life stage or phenophase can be expressed as a constant having physiological units such as day-degrees (Manetsch 1976; Welch et al. 1978; Gutierrez et al. 1984b). As a result, the population and its cohorts travel on a physiological time horizon that substitutes for calendar time (Gilbert et al. 1976; Gutierrez and Baumgärtner 1984b, Gutierrez et al. 1984b). This constant can be modified to account for nutritional effects (Gutierrez et al. 1984a, 1988b). In the second approach, D_T becomes time-varying $[D\{T(t)\}]$ to account for the effects of temperature (T) and other environmental variables on transit times.

The dynamics of N are described by [A.II.11], where $k = D^2/\text{var}$, var is the variance of developmental times, ageing occurs via flow rates r_{i-1} from N_{i-1} to N_i, births enter the population via $r_0(t) = x(t)$, and deaths at maximum age exit via $y(t)$. (We note that one could use numbers just as easily as rates.)

$$\frac{dN_1}{dt} = x(t) - r_1(t)$$

$$\frac{dN_2}{dt} = r_1(t) - r_2(t)$$

$$\vdots \qquad\qquad\qquad\qquad \text{[A.II.11]}$$

$$\frac{dN_k}{dt} = r_{k-1}(t) - y(t)$$

By using Euler integration and under the assumption $r_i = N_i \cdot k/D$, we can approximate [A.II.11] over the interval $[t, t + \Delta t]$ as follows:

$$r_1(t + \Delta t) = r_1(t) + \Delta t \left(\frac{k}{D}\right) [x(t) - r_1(t)]$$

$$r_2(t + \Delta t) = r_2(t) + \Delta t \left(\frac{k}{D}\right) [r_1(t) - r_2(t)]$$

$$\vdots \qquad\qquad\qquad\qquad\qquad \text{[A.II.12]}$$

$$y(t + \Delta t) = y(t) + \Delta t \left(\frac{k}{D}\right) [r_{k-1}(t) - y(t)]$$

where r_i is the flow rate (i.e., aging rate) from the i to the $i + 1$ cohort and $N_i(t)$ is the number in the ith cohort. The number in the total population equals

$$N(t) = \sum_{i=1}^{k} N_i(t)$$

If the age specific mortalities are zero and if $k = 1$, the developmental times of a cohort would have a geometric distribution, otherwise larger values of k would produce a distribution of development times characterized by a delay to first emergence and a variance about the mean emergence time. Note that the variance goes to zero as $k \to \infty$. The resulting distribution of developmental times for several values of k are shown below (see Vansickle 1977).

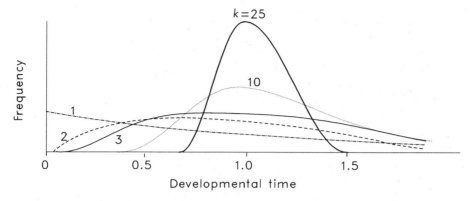

Frequency distributions of developmental times in the absence of attrition using different values of k.

ADDING ATTRITION. If we are modeling the number dynamics of a population with age-specific time-varying mortality $[\mu_i(\cdot)]$ to cohorts N_i, the dynamics are described by [A.II.13].

$$\frac{dN_1}{dt} = x(t) - r_1(t) - \mu_1(t, \cdot)N_1(t)$$

$$\frac{dN_2}{dt} = r_1(t) - r_2(t) - \mu_2(t, \cdot)N_2(t)$$

$$\vdots$$

[A.II.13]

$$\frac{dN_k}{dt} = r_{k-1}(t) - y(t) - \mu_k(t, \cdot)N_k(t)$$

The addition of time-varying mortality is an important component introducing realism into our models. The instantaneous solution for flow rates is as follows (Vansickle 1977)

$$\frac{dr_i}{dt} = \frac{k}{D}\left\{ r_{i-1}(t) - \left[1 + \mu_i(t, \cdot)\frac{D}{k} \right] r_i(t) \right\}$$ [A.II.14]

where the net proportion age-specific mortality (i.e. the net proportion of death, immigration, and emigration) during dt equals $-\infty < \mu_i \leq 1$ and individuals surviving to maximum age exit as $r_k(t) = y(t)$. Note that $\mu \to \infty$ if massive immigration occurs. Of course, model [A.II.13] requires initial conditions to insure the uniqueness of the solution.

Generalizations of Deterministic Age Structure Models to Several Variables. Leslie's model has been used by many investigators to include attributes other than time and age. Slobodkin's (1961) work on *Daphnia obtusa* suggested that in addition to age, size should also be taken into account in defining classes of physiologically identical organisms. This is an important contribution to the development of realism in such a model and is an important focus of this book.

Discrete Form: Slobodkin's Model. Slobodkin's model assumes that organisms of the same age and size are physiologically identical. Similar to Leslie's model, at each time t, entities $N_{i,j}$ are defined to be the number of individuals in size bracket i and age bracket j. Again we assume that $\Delta t = \Delta a$, so that with each increment of time, the organisms are moved one age category ahead. However, they may or may not move ahead one size category. Here we assume that the scales are so chosen so that it is impossible for any individual to increase more than one size category within one increment of time. The decision whether or not to move $N_{i,j}$ to $N_{i,j+1}$ or to $N_{i+1,j+1}$ may depend on many parameters such as the time of the season, the age and the size of the organisms, the size of the population, and the availability of resources. Slobodkin made the simplification that individuals in one category either all grow to the next size category or all remain in the same size bracket. Similar simplification is made at death, that is, either none of the individuals in $N_{i,j}$ die or all of them die.

Algebraically, we have for one increment of time, using time as a superscript

$$N_{i,j}^{t+\Delta t} = \alpha N_{i-1,j-1}^{t} + \beta N_{i,j-1}^{t} \qquad \text{[A.II.15]}$$

where α and β take on the value 0 or 1. If there is no death in $N_{i-1,j-1}$, and it moves in both age and size, then $\alpha = 1$, otherwise $\alpha = 0$. If there is no death in $N_{i,j-1}$, and it moves only in age and not in size then $\beta = 1$, otherwise $\beta = 0$. Slobodkin's simplification can easily be relaxed so that in one category, not all individuals die or not die, and grow or not grow. In fact, the parameters α and β in equation [A.II.15] can very well be functions of time, age, size, and so on, taking values between 0 and 1.

From [A.II.15], we have seen that the category $N_{i,j}[x]$ at time $t + \Delta t$ depends on the values of the categories $N_{i-1,j-1}^{t+\Delta t}[X]$ and $N_{1,j-1}^{t}[X]$ at time t. Graphically, we have the point dependence as illustrated below.

Continuous Form: Sinko–Streifer Equation (1967, 1969). Let $N(t, a, m)$ be the population density function depending on t = time, a = age, and m = mass (or size). Many biological entities are determined by the population density function $N(t, a, m)$. For example,

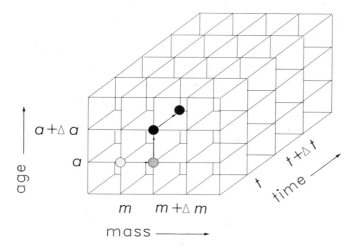

1. The number of organisms between ages (a_0, a_1) and sizes (m_0, m_1) is

$$\int_{a_0}^{a_1} \int_{m_0}^{m_1} N(t, a, m) \, dm \, da$$

2. The total biomass (in units of m) of organisms at time t is

$$\int_0^\infty \int_0^\infty mN(t, a, m) \, dm \, da$$

If we assume that one unit of increase in time gives one unit of increase in age, then $N(t, a, m)$ satisfies

$$\frac{\partial N}{\partial t} + \frac{\partial N}{\partial a} + \frac{\partial [g(\cdot)N(t, a, m)]}{\partial m} = -\mu(\cdot)N(t, a, m) \qquad \text{[A.II.16]}$$

where $g(\cdot)$ and $\mu(\cdot)$ are the growth and the death rates, respectively. They may depend on various parameters dictated by the particular problem. Equation [A.II.16] needs two additional conditions to complete its description of the population density function:

1. The age–size distribution of the population at time 0 is $N(0, a, m)$
2. The size density of newborn is $N(t, 0, m)$.

We denote $N(0, a, m)$ and $N(t, 0, m)$ by $\alpha(a, m)$ and $\beta(t, m)$, respectively. Not all equations of the form of [A.II.16] have analytic solutions. If g is a function of t and a only and μ is independent of $N(t, a, m)$, then [A.II.16] has analytic solutions given by

$$N(t, a, m) = \alpha\left[a - t, m - \int_0^t g(t', t' + a - t)\, dt'\right]$$

$$\cdot \exp\left\{-\int_0^t\left[\mu(t', t' + a - t, m)\right.\right.$$

$$\left.\left. - \int_{t'}^t g(t'', t'' + a - t)\, dt''\right] dt'\right\}$$

for $a > t$, and

$$= \beta\left[t - a, m - \int_{t-a}^t g(t', t' - t + a)\, dt'\right]$$

$$\cdot \exp\left\{-\int_{t-a}^t\left[\mu(t', t' - t + a, m)\right.\right.$$

$$\left.\left. - \int_{t'}^t g(t'', t'' - t + a)\, dt''\right] dt'\right\}$$

for $a < t$.

Whether equation [A.II.16] can be solved analytically or not depends on the complexity of g, μ, and β. To solve equation [A.II.16] numerically, the following discretization scheme is used,

$\dfrac{\partial N}{\partial t}$ is approximated by $\dfrac{N(t + \Delta t, a + \Delta a, m) - N(t, a + \Delta a, m)}{\Delta t}$

$\dfrac{\partial N}{\partial a}$ is approximated by $\dfrac{N(t, a + \Delta a, m) - N(t, a, m)}{\Delta a}$

and

$\dfrac{\partial N}{\partial m}$ is approximated by $\dfrac{N(t, a, m) - N(t, a, m - \Delta m)}{\Delta m}$

If we again assume that $\Delta t = \Delta a$ and that one increment in age causes an increase of no more than one size category, then [A.II.16] becomes

$$N(t + \Delta t, a + \Delta a, m) = \left\{1.0 - \frac{g(t, a) \cdot \Delta t}{\Delta m} - \mu(\cdot)\Delta t\right\} N(t, a, m)$$

$$+ \frac{g(t, a)}{\Delta t} N(t, a, m - \Delta m) \qquad \text{[A.II.17]}$$

Equation [A.II.17] is equivalent to [A.II.16] except α and β now take more complicated functional forms. The point dependence graph above also applies to [A.II.17]. One should note that the approximations of the partial derivatives are not arbitrary. The scheme should be so designed that it is reasonable for the problem under consideration. This is further discussed in numerical analysis (e.g., Richtmyer and Morton 1967; Courant and Hilbert 1962) and we will not go into the details here.

We have discussed in detail balance equations of three independent variables. The number of variables does not have to be limited to any quantity. In general, suppose there are k physical characteristics $m_1, m_2, \ldots, m_k$, then the balance equation is ($da/dt = 1$)

$$\frac{\partial N}{\partial t} + \frac{\partial N}{\partial a} + \sum_{i=1}^{j} \frac{\partial}{\partial m_j} g_j(\cdot)N = -\mu(\cdot)N \qquad \text{[A.II.18]}$$

where $g(\cdot) = dm_j/dt =$ the growth rate of characteristic m_j. Equation [A.II.18] is a generalization of [A.II.10] and [A.II.16].

Interaction of Several Species

So far, we have discussed population models for single species. Most populations do not, however, live in isolated environments. Nature is full of biological association, which can take many forms. Two species may compete for a limited environment whether it be food or space or other requisites. One species may eat the other species so that the predacious species adversely affects the increase of the other. The second species may live on the waste products of the first, doing the first species neither direct good nor harm. If the host dies as a result of parasitism, then this relationship is more like that of predator-prey except a parasite normally kills only one host during its lifetime. The host frequently does not disappear immediately but in some future time. Because of the complexity of the relationships, the construction of these models is not easy. The difficulty lies mainly in the formation of parameter functions (e.g., functional response models), which accurately describe the interactions under consideration. The stability properties of such systems are examined in Appendix III. Here we introduce some mathematical notions.

Simple Models: One Independent Variable. As we have seen in the previous sections, models with only one independent variable (time) are in general not realistic. However, the analysis is simplest in these cases, so we now study them in detail.

Linear Models. We consider the interaction of two species. Let $N_1(t)$ and $N_2(t)$ be the number of individuals at time t of species 1 and 2, respectively. Suppose

$N_1(t)$ and $N_2(t)$ satisfy the following system of ordinary differential equations:

$$\frac{dN_1}{dt} = a_{11}N_1(t) + a_{12}N_2(t)$$

$$\frac{dN_2}{dt} = a_{21}N_1(t) + a_{22}N_2(t) \qquad\qquad [\text{A.II.19}]$$

In matrix notation, [A.II.19] is equivalent to

$$\begin{pmatrix} \dfrac{dN_1(t)}{dt} \\ \dfrac{dN_2(t)}{dt} \end{pmatrix} = \begin{pmatrix} a_{11} & a_{12} \\ a_{21} & a_{22} \end{pmatrix} \begin{pmatrix} N_1(t) \\ N_2(t) \end{pmatrix} \qquad\qquad [\text{A.II.19a}]$$

or $\dot{N}(t) = AN(t)$.

Consider the case that matrix A is constant. If $a_{12} = a_{21} = 0$, then the two species are independent of each other. Here, we assume the contrary and note that the nature of the associations dictates the nature of the parameters in [A.II.19].

Example a. Prey–Predator and Host–Parasite Relationships. Since the second species harms the first species, therefore, the more second species we have, the less first species we have. Hence, we have $a_{12} < 0$. Conversely, the first species is advantageous to the second, so $a_{21} > 0$. For pure predation and parasitism relationships, the second species will go to extinction in the absence of the first, therefore $a_{22} < 0$.

Example b. Competition. Since the abundance of one species is to the disadvantage of the other, we have $a_{12} < 0$ and $a_{21} < 0$.

Example c. Scavenging. Suppose the second species is the scavenger, then $a_{21} > 0$. Since species 1 is not affected by the presence or absence of species 2, $a_{12} = 0$.

The solutions of [A.II.19a] can be determined and used to analyze the properties of the model. This topic is covered in great detail in Appendix III. Let λ be a real number. We wish to determine if there exist one or more column vectors such that

$$\begin{pmatrix} a_{11} & a_{12} \\ a_{21} & a_{22} \end{pmatrix} \begin{pmatrix} N_1 \\ N_2 \end{pmatrix} = \lambda \begin{pmatrix} N_1 \\ N_2 \end{pmatrix}$$

Rearranging the terms yields

$$\begin{pmatrix} a_{11} & a_{12} \\ a_{21} & a_{22} \end{pmatrix} \begin{pmatrix} N_1 \\ N_2 \end{pmatrix} - \lambda \begin{pmatrix} N_1 \\ N_2 \end{pmatrix} = 0$$

and multiplying λ by the identity matrix gives

$$\begin{pmatrix} a_{11} & a_{12} \\ a_{21} & a_{22} \end{pmatrix} \begin{pmatrix} N_1 \\ N_2 \end{pmatrix} - \lambda \begin{pmatrix} 1 & 0 \\ 0 & 1 \end{pmatrix} \begin{pmatrix} N_1 \\ N_2 \end{pmatrix} = 0$$

Factoring out the column vector (N_1, N_2) simplifies the expression.

$$\begin{pmatrix} a_{11} - \lambda & a_{12} \\ a_{21} & a_{22} - \lambda \end{pmatrix} \begin{pmatrix} N_1 \\ N_2 \end{pmatrix} = 0$$

The λ that satifies the equation may be obtained by taking the determinants of the matrix as follows:

$$\lambda^2 - (a_{11} + a_{22})\lambda + a_{11}a_{22} - a_{12}a_{21} = 0 \qquad \text{[A.II.20]}$$

where λ_1 and λ_2 are roots of the equation. If the roots are real, the solutions to [A.II.19] are

$$N_1(t) = c_1 e^{\lambda_1 t} + c_2 e^{\lambda_2 t}$$

$$N_2(t) = d_1 e^{\lambda_1 t} + d_2 e^{\lambda_2 t}$$

where c_1, c_2, d_1, and d_2 are positive constants (see Appendix III).

The two populations may go to extinction, grow, or oscillate simultaneously depending on the coefficients in [A.II.20]. In [A.II.19a], the matrix **A** need not be constant, it can be time dependent, but in this case the solution cannot be explicitly solved unless time dependence is periodic. We can now apply our solution to a simple model.

Nonlinear Models. The general nonlinear model is of the form:

$$\frac{dN_1}{dt} = F(N_1, N_2)$$

$$\frac{dN_2}{dt} = G(N_1, N_2)$$

for two interacting species. We look at a special case

$$\frac{dN_1}{dt} \; a_1 N_1(t) \; - \; a_{11} N_1^2(t) \; - \; a_{12} N_1(t) N_2(t)$$

$$\frac{dN_2}{dt} \; a_2 N_2(t) \; - \; a_{22} N_2^2(t) \; - \; a_{21} N_2(t) N_1(t) \qquad \text{[A.II.21]}$$

where

$$F(N_1, N_2) \; = \; r_1 N_1 \frac{K_1 - N_1 - \alpha N_2}{K_1}$$

and

$$G(N_1, N_2) \; = \; r_2 N_2 \frac{K_2 - N_2 - \beta N_1}{K_2} \qquad \text{[A.II.22]}$$

where: $r_1 = \alpha_1$, $r_2 = \alpha_2$, $K_1 = a_1/a_{11}$, $K_2 = a_2/a_{22}$, $\alpha = a_{12}/a_{11}$, and $\beta = a_{21}/a_{22}$. r_1 and r_2 are the intrinsic rates of increase of species 1 and 2, and K_1 and K_2 are the saturation levels of species 1 and 2, respectively.

If the association between two species is competition, then α and β are the competition coefficients. The interaction of two competing species may result in species coexistence or the exclusion of one:

Case 1: If $\alpha < \dfrac{K_1}{K_2}$ and $\beta > \dfrac{K_2}{K_1}$ then species 1 wins.

Case 2. If $\alpha > \dfrac{K_1}{K_2}$ and $\beta < \dfrac{K_2}{K_1}$ then species 2 wins.

Case 3. If $\alpha > \dfrac{K_1}{K_2}$ and $\beta > \dfrac{K_2}{K_1}$ then survival depends on the initial population size.

Case 4. If $\alpha < \dfrac{K_1}{K_2}$ and $\beta < \dfrac{K_2}{K_1}$ then species 1 and 2 coexist.

The properties of the two species interaction model may be determined graphically. First set

$$\frac{dN_1}{dt} = \frac{dN_2}{dt} = 0$$

and determine the zero isocline for each species. Trivial solutions are when r_1, N_1 or both are zero, hence the problem reduces to the solution of

$$K_1 - N_1 - \alpha N_2 = 0$$

which is

$$N_2 = \frac{K_1}{\alpha} - \frac{N_1}{\alpha}$$

The zero isocline for N_2 is the following:

$$N_1 = \frac{K_2}{\beta} - \frac{N_2}{\beta}$$

These isoclines are illustrated in the figure below where the vector field has N_1 and N_2 components. Note that a simple way to evaluate the vectors is to remember that the vector for the N_1 population moves in the direction of K_1 from either above or below, and similarly that for N_2 moves in the direction of K_2 from above and below. The resultant movement of the sum of the vectors are shown in the figures.

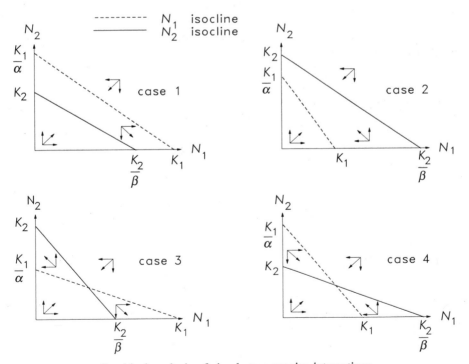

Graphical analysis of simple two species interactions.

Complex Models: Several Independent Variables. We have concerned ourselves with two interacting populations only. In general, if we have m interacting species, the model with one independent variable is of the form

$$\frac{dN_1}{dt} = F_i(N_1, N_2, \ldots, N_m) \qquad i = 1, 2, \ldots, m \qquad [\text{A.II.23}]$$

where the determination of the form of F_i depends on the problem under consideration. It is conceivable that as the number of species increase, the formulation of the association parameter functions becomes increasingly difficult.

Suppose we have m interacting species. The general form of the mathematical model for n independent variables is as follows:

$$\frac{\partial N_i}{\partial t} + \frac{\partial N_i}{\partial a}\frac{da}{dt} + \sum_{j=1}^{n-2} \frac{\partial}{\partial s_j} (g_j N_i) = F_i(N_1 N_2, \ldots, N_m) \qquad [\text{A.II.24}]$$

for $i = 1, 2, \ldots, m$

Models with this complexity have not been used heavily in the past. Chapter 10 introduces in some detail the development of field models of the form [A.II.24] for plants and animals. Models with distributed developmental times may also be used without adversely increasing the complexity of the system. In practice, however, neither model [A.II.24] or the distributed delay models [A.II.13] are likely to yield analytic solutions.

Appendix III

Introduction to Stability Analyses of Population Models

During the mid-1920s, Lotka (a physical chemist and demographer) and Volterra (a mathematician) introduced some of the rigor of the physical sciences to population ecology. Lotka applied the notions of what we now call systems theory to the analysis of intra- and interspecific interactions in population ecology. Their objective was to model the interrelationships between the component parts of a predator–prey system and to analyze the system's behavior mathematically. This work anticipated Ludwig von Bertalanffy's development after World War II of general systems theory (see von Bertalanffy 1962). May (1973) popularized the use of rigorous analyses in animal ecology and, particularly with colleagues, he sought to examine the factors responsible for the apparent stability of animal populations in nature. In this appendix, we approach the subject and its applications to population ecology in an analytical yet intuitive manner. This appendix is modeled after the excellent appendix written by Roughgarden (1979).

EQUILIBRIUM LEVEL AND SYSTEM STABILITY

In biological terms, the stability of a population (i.e., a system) is its propensity to fluctuate about some average density, neither increasing without bounds nor going to extinction. There may be one or many interacting species in the system. More precisely, the stability of a system is characterized as its ability to return to equilibrium density (N^*) after perturbation. In general, there are two salient questions: What is the equilibrium population density for each species in a system, and is the system stable? An allied concept not covered in detail here is Holling's (1973) concept of resilience, which is defined as the

237

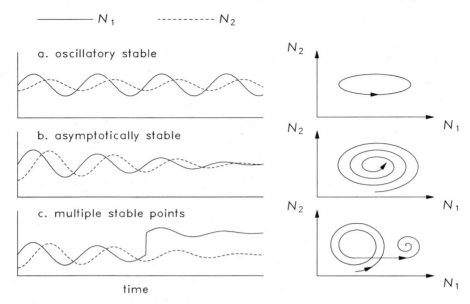

Figure A.III.1. *Examples of predator–prey interactions and their stability characteristics. The phase plane diagrams are shown on the left-hand side.*

ability of a system to return to equilibrium, though not necessarily the same level, after perturbation (i.e., multiple equilibria).

The concepts of population dynamics and of stability are illustrated in Figure A.III.1A–C using hypothetical predator–prey interactions and the phase plots of the dynamics (i.e., time series plot of predator density on prey density). Some systems may have no discernible pattern of interaction and are often presumed to be chaotic (not illustrated). To interpret these figures with respects to one species (say the prey), think of a ball perfectly balanced on a wire, which if perturbed a distance η along the axis of the wire, may be unstable (Fig. A.III.2A) and grow increasingly distant from N^*; may oscillate (Fig. A.II.2b), may be in a very stable node (Fig. A.III.2c), or may be displaced to another stable level (2d, i.e., resilience, multiple equilibria). A system may exhibit different stability behaviors depending on the parameters (Fig. A.III.3).

POPULATION MODELS

Population models are used to describe the dynamics of populations, and they may be cast as differential or difference equations (see the text and Appendix I). An example of a differential equation model of a single species is

$$\frac{dN}{dt} = f(N)$$

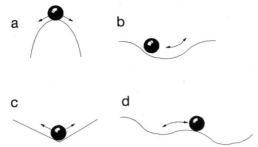

Figure A.III.2. *Types of stability: (a) unstable, (b) oscillatory stable, (c) asymptotically stable (cf. Berryman 1981), and (d) resilient (Holling 1973) and multiple equilibria.*

The equilibria are determined by finding the values $N = N^*$ that satisfy

$$f(N^*) = 0$$

For a discrete model such as

$$N(t + 1) = f(N(t))$$

the equilibria are determined by finding N^* such that

$$N^* = f(N^*)$$

We may determine if a system composed of one (or several) species is stable using graphically or analytically methods. In these analyses, one wishes to know the behavior of the system when one (or more) of the component species

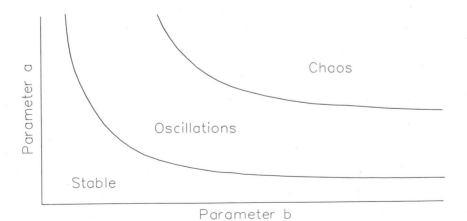

Figure A.III.3. *Hypothetical stability characteristics in parameter space (a,b).*

is perturbed a small amount η from the equilibrium level (say $N*$ to $N* + \eta$): specifically, does the system return to $N*$ (i.e., $|\eta(t)| \to 0$) making the system stable or does the perturbation increase possibly making the system chaotic.

The steps used to find the stability properties of differential or difference models using graphical or analytical methods are as follow:

Step 1: Determine the equilibria densities of each species (i.e., the condition of no change) defined for differential equations as $dN/dt = 0$ and for difference equations as $N(t + 1)/N(t) = 1$.

Step 2: Determine if the system is stable using graphical or analytical methods.

Stability Properties of Differential Equation Models

Below, we investigate the stability of equilibria for a single species [A.III.1a]

$$\frac{dN}{dt} = f(N) \qquad [A.III.1]$$

and for two species [A.III.2]

$$\frac{dN_i}{dt} = f(N_1, N_2) \qquad [A.III.2]$$

Graphical Analyses

Single Species Models. Assume that a population's growth rate is logistic and may be described by

$$\frac{dN}{dt} = rN \left(1 - \frac{N}{K} \right) \qquad [A.III.3]$$

where r is the per capita population growth rate and K is the carrying capacity of the environment ($K = N_{max}$).

Step 1: Assume that $r > 0$ and $K > 0$, hence to determine the equilibrium population level, we solve $dN/dt = 0$ for values of N that satisfy this equality. Solutions satisfying this equilibrium condition are $N = 0$ and $N = K$. Only the latter is nontrivial.

Step 2:

$$\frac{dN}{dt} < 0 \quad \text{for} \quad N > K \quad \text{and} \quad \frac{dN}{dt} > 0 \quad \text{and} \quad 0 < N < K,$$

hence, all solutions to [A.III.3] approach K. Graphically, we can illustrate this in one dimension as follows:

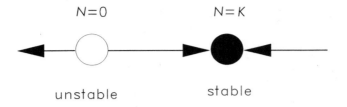

$$N=0 \qquad\qquad N=K$$

unstable   stable

Two Competing Species. The procedures for analyzing the stability properties of two competing species N_1 and N_2 are similar to the single species case. Suppose the logistic growth of two competing species N_1 and N_2 are described by

$$\frac{dN_1}{dt} = r_1 N_1 \left[1 - \frac{(N_1 + \alpha N_2)}{K_1} \right]$$

and

$$\frac{dN_2}{dt} = r_2 N_2 \left[1 - \frac{(N_2 + \beta N_1)}{K_2} \right]$$

where r_1 and r_2 are the positive per capita growth rates of species N_1 and N_2, α and β are the effects of N_1 on N_2 and N_2 on N_1, respectively, and K_1 and K_2 are the environmental carrying capacities for N_1 and N_2 (see Appendix II).

Step 1: Solve for $dN_1/dt = dN_2/dt = 0$. For N_1,

$$r_1 N_1 \left(1 - \frac{N_1 + \alpha N_2}{K_1} \right) = 0$$

has solutions $\begin{cases} N_1 = 0 \text{ (trivial) and} \\ 1 - (N_1 + \alpha N_2)/K_1 = 0 \text{ (nontrivial)} \end{cases}$

Rearrange the terms of the nontrivial solution yields

$$N_2 = \frac{K_1}{\alpha} - \frac{N_1}{\alpha}$$

This line is called the N_1 zero isocline (all N_1 and N_2 satisfying $dN_1/dt = 0$). The above steps are repeated for population N_2 to determine its isocline.

Step 2: Plot the N_1 and N_2 isoclines (Fig. A.III.4). Note that $-1/\alpha$ and $-\beta$ are the slopes and K_1/α and K_2 are the intercepts. We need to determine roughly, the direction of the vectors (dN_1/dt, dN_2/dt) for all values of N_1 and N_2 (i.e., the vector arrows in Fig. A.III.4).

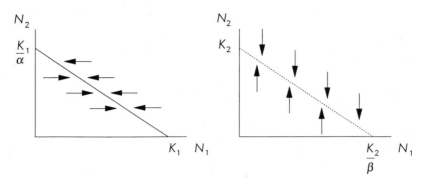

Figure A.III.4. *Movement of population vectors with regards to the N_1 (——) and N_2 (-------) isoclines.*

We get this basic information by noting that

$$\frac{dN_i}{dt} = \begin{cases} < 0 \text{ if } (N_1, N_2) \text{ lies above the } N_i \text{ isocline} \\ = 0 \text{ if } (N_1, N_2) \text{ lies on the } N_i \text{ isocline} \\ > 0 \text{ if } (N_1, N_2) \text{ lies below the } N_i \text{ isocline} \end{cases}$$

$$i = 1, 2$$

Hence, from any starting point (N_1, N_2) in the phase plane (Fig. A.III.4), the direction of change for each species is along it's axis toward the isocline either from above or below. For interacting populations, the direction is the resultant or sum of the vectors (Fig. A.III.5). The system in Fig. A.III.5a is stable because there is a common point of convergence (other possible combinations of interactions are given in Appendix II), but what occurs in the arbitrary system illustrated in Fig. A.III.5b? Finally, a word of caution is in order—graphical

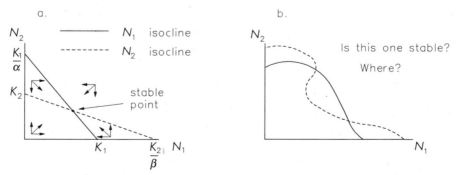

Figure A.III.5. *Graphical analyses of two species systems—two examples. Note that the vectors in A are derived from analyses such as those depicted in Figure 4.*

analyses may provide inconclusive results in even apparently simple cases and may require further analysis.

Analytical Methods to Determine the Stability of Equilibria of Differential Equation Models. The analysis requires that the dynamics model for each species [e.g., $dN/dt = f(N)$] be continuously differentiable with respect to N.

Single Species. Consider $dN/dt = f(N)$. By definition, the system is at equilibrium when $dN/dt = 0$ or $f(N^*) = 0$. However, suppose that we have a solution $N(t)$ with $N_0 = N^* + \eta^*$ (Fig. A.III.6, point a). What we wish to know is whether the system returns to N^* (i.e., $|N(t) - N^*| \to 0$) (Fig. A.III.6), or moves away from it

We need to evaluate η as it changes over time after the initial perturbation. If $\eta(t) = N(t) - N^*$, notice that

$$\frac{d\eta(t)}{dt} = \frac{d}{dt}\,[N(t) - N^*] = \frac{dN(t)}{dt} - 0 = f(N(t))$$

Using Taylor's theorem we approximate $f(\eta + N^*)$ by its linear part. As $f(N^*) = 0$, we have

$$\frac{d\eta}{dt} \approx f(\eta + N^*) \approx f(N^*) + f'(N^*)\eta = f'(N^*)\eta$$

Note that $f(N^*) = 0$, and if η is sufficiently small, all higher order terms may be ignored giving us a linear approximation of f near N^*. Therefore,

$$f(\eta + N^*) \approx \frac{d\eta}{dt} \approx f'(N^*)\eta$$

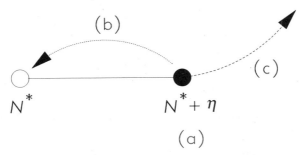

Figure A.III.6. *System stability: (a) displacement of a population from equilibrium (N^*) a distance η, (b) the system is stable and the population returns to N^*, (c) the system is unstable the system and the population grows away from N^*.*

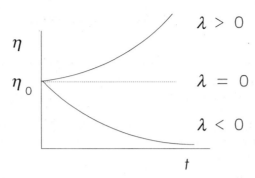

Figure A.III.7. The behavior of η over time (see above).

Let $\lambda = f'(N^*)$, then $d\eta/dt \approx \tilde{\lambda}\eta$, and the solution is

$$\eta_t = \eta_o e^{\lambda t}$$

where $\eta_0 = N_0 - N^*$ is the initial perturbation from N^*. The behavior of η over time is shown in Fig. A.III.7 for different λ: Thus, we may conclude for $dN/dt = f(N)$

1. If $\lambda < 0$, η_0 decays exponentially to zero and the system is stable at N^*.
2. If $\lambda = 0$, no conclusion can be made.
3. If $\lambda > 0$, the system is unstable.

For the logistic model, $f(N) = r_N(1 - N/K)$, and 0 and K are the equilibria. Since $f'(K) = -r < 0$ and $f'(0) = r > 0$, we see immediately that $N^* = 0$ is unstable and $N^* = K$ is stable.

Analysis of a Two-Species System. The analysis for a two-species system is very similar to that outlined for a one-species system—it is just a bit more complicated. Assume a system

$$\frac{dN_1}{dt} = f(N_1, N_2)$$

$$\frac{dN_2}{dt} = g(N_1, N_2)$$

[A.III.4]

We further assume that

1. $N_1(t)$, $N_2(t)$ is a solution to [A.III.4]. Let $\eta_1(t) = N_1(t) - N_1^*$ and $\eta_2(t) = N_2(t) - N_2^*$, respectively.
2. Again we see that

$$\frac{d\eta_1}{dt} = \frac{dN_1}{dt} = f(N_1, N_2) = f(\eta_1 + N_1, \eta_2 + N_2)$$

$$\frac{d\eta_2}{dt} = \frac{dN_2}{dt} = g(N_1, N_2) = g(\eta_1 + N_1, \eta_2 + N_2)$$

3. Via Taylor expansions, we get a linear approximation of functions f

$$\frac{d\eta_1}{dt} \approx f(N_1^*, N_2^*) + \frac{\partial f(N_1^*, N_2^*)}{\partial \eta_1} \eta_1 + \frac{\partial f(N_1^*, N_2^*)}{\partial \eta_2} \eta_2$$

and

$$\frac{d\eta_2}{dt} \approx g(N_1^*, N_2^*) + \frac{\partial g(N_1^*, N_2^*)}{\partial \eta_1} \eta_1 + \frac{\partial g(N_1^*, N_2^*)}{\partial \eta_2} \eta_2$$

4. Let

$$a = \frac{\partial f(N_1^*, N_2^*)}{\partial \eta_1} \qquad b = \frac{\partial f(N_1^*, N_2^*)}{\partial \eta_2} \qquad c = \frac{\partial g(N_1^*, N_2^*)}{\partial \eta_1}$$

$$\text{and} \qquad d = \frac{\partial g(N_1^*, N_2^*)}{\partial \eta_2}$$

and

$$f(N_1^*, N_2^*) = g(N_1^*, N_2^*) = 0$$

5. We can rewrite step 3 as follows

$$\frac{d\eta_1}{dt} = a\eta_1 + b\eta_2$$

$$\frac{d\eta_2}{dt} = c\eta_1 + d\eta_2$$

Recall from linear algebra (see Appendix I for a review) that the left hand side of the step 5 can be written in vector form:

$$\begin{pmatrix} \dfrac{d\eta_1}{dt} \\ \dfrac{d\eta_2}{dt} \end{pmatrix} = \begin{pmatrix} \dot{\eta}_1 \\ \dot{\eta}_2 \end{pmatrix} = \dot{\eta},$$

and the matrix of the partial derivatives (see step 4) (i.e., the Jacobian matrix) may be written as

$$\mathbf{A} = \begin{pmatrix} a & b \\ c & d \end{pmatrix}$$

6. The problem is to solve step 3 for sufficiently small $\eta_1(0)$ and $\eta_2(0)$. To do this we begin by finding the eigenvalues of the matrix **A** [i.e., the numbers (λ) that determine if the populations return to equilibrium or not]. This means solving the algebraic equation

$$\begin{pmatrix} a & b \\ c & d \end{pmatrix} \begin{pmatrix} \eta_1 \\ \eta_2 \end{pmatrix} = \lambda \begin{pmatrix} \eta_1 \\ \eta_2 \end{pmatrix}$$

Rearranging

$$A \begin{pmatrix} \eta_1 \\ \eta_2 \end{pmatrix} - \lambda \begin{pmatrix} \eta_1 \\ \eta_2 \end{pmatrix} = 0$$

Multiplying by an identity matrix (I),

$$I = \begin{pmatrix} 1 & 0 \\ 0 & 1 \end{pmatrix}$$

as follows provides a method for finding the roots (i.e., the λ values).

$$\begin{pmatrix} a & b \\ c & d \end{pmatrix} \cdot \begin{pmatrix} \eta_1 \\ \eta_2 \end{pmatrix} - \lambda \begin{pmatrix} 1 & 0 \\ 0 & 1 \end{pmatrix} \begin{pmatrix} \eta_1 \\ \eta_2 \end{pmatrix} = 0$$

This expression is equivalent to

$$\begin{pmatrix} a - \lambda & b \\ c & d - \lambda \end{pmatrix} \cdot \begin{pmatrix} \eta_1 \\ \eta_2 \end{pmatrix} = 0$$

This equation has nontrivial solutions if and only if the determinant of

$$\begin{pmatrix} a & b \\ c & d \end{pmatrix} \quad \text{is} \quad 0$$

7. We find the determinants of the matrix

$$\det \begin{pmatrix} a - \lambda & b \\ c & d - \lambda \end{pmatrix} = (a - \lambda)(d - \lambda) - bc$$

$$= \lambda^2 - (a + d)\lambda + ad - bc$$

and setting the equation to 0 we solve for the roots λ using the quadratic formula

$$\lambda = \frac{(a + d) \pm \sqrt{(a + d)^2 - 4(ad - bc)}}{2}$$

This yields two roots λ_1 and λ_2 (i.e., eigenvalues), that can be used to find the eigenvectors (η_1, η_2).

The general solution for linear equations (e.g., steps 3 and 5) is

$$c_1 \begin{pmatrix} \eta_1 \\ \eta_2 \end{pmatrix} e^{\lambda_1 t} + c_2 \begin{pmatrix} \eta_1 \\ \eta_2 \end{pmatrix} e^{\lambda_2 t}.$$

Different solutions of λ_1 and λ_2 yield a wide array of outcomes and readers are referred to Lotka (1925, page 148) or the appendix in Roughgarden (1979) for historical details and for a complete list of possible outcomes. Stability is assumed if either the roots λ_1 and λ_2 are real and negative, or the roots are complex numbers (i.e., $\lambda_i = \alpha_i \pm \beta_i$) and the α_i are negative.

Stability Properties of Difference Equation Models

We shall reverse the order of our discussion in order to show its parallel to the previous section on differential models. We first review analytical methods for determining the stability of difference models, and then present graphical methods.

Analytical Methods

Single Species. Assume that $N_{t+1} = g(N_t)$, and that $dg(N)/dN$ exists. Hence, to examine the stability properties of the system, we use the following steps:

1. The system is at equilibrium when $N_{t+1} = N_t$ or $g(N) = N^*$.
2. Examine the system with initial conditions near N^*, and $\eta_t = N_t - N^*$. If $|\eta| \to 0$, then the system is stable.
3. We replace the N values in $N_{t+1} = g(N_t)$ with $\eta + N^*$,

$$\eta_{t+1} + N^* = g(\eta_t + N^*)$$

4. A Taylor's expansion of g near N^* yields

$$\eta_{t+1} + N^* \approx g(N^*) + g'(N^*)\eta_t$$

5. Because $g(N^*) = N^*$, we can make appropriate substitution in step 4, yielding the linear approximation

$$\eta_{t+1} + N^* \approx N^* + g'(N^*)\eta_t$$

and rearranging terms we get

$$\eta_{t+1} \approx g'(N^*)\eta_t$$

Therefore, if we replace $g'(N^*)$ by c

If $|c| < 1$, $|\eta| \to 0$, the equilibrium is stable

If $|c| > 1$, then equilibrium is unstable.

If $c < 0$, the sign of $(g^n)'(N^*)$ changes each time step and we get oscillatory stability and instability

Two Species

$$N_{t+1} = f(N_t, P_t)$$

$$P_{t+1} = g(N_t, P_t)$$

We find the equilibria by (N_t^*, P_t^*) by solving the equations

$$N = f(N, P) \text{ and } P = g(N, P)$$

To determine the stability of the system, consider (N_0, P_0) as initial conditions sufficiently close to (N^*, P^*). Let $\eta_t = N_t - N^*$ and $\rho_t = P_t - P^*$. To show the stability of (N^*, P^*), we need to show that $|\eta_t, \rho_t| \to 0$ as $t \to \infty$. Again, we use the Taylor series to approximate

$$\eta_{t+1} + N^* = f(\eta_t + N^*, \rho_t + P^*)$$

$$\approx f(N^*, P^*) + \frac{\partial f(N^*, P^*)}{\partial N} \eta_t + \frac{\partial f(N^*, P^*)}{\partial P} \rho_t$$

and

$$\rho_{t+1} + P^* = g(\eta_t + N^*, \rho_t + P^*)$$

$$\approx g(N^*, P^*) + \frac{\partial g(N^*, P^*)}{\partial N} \eta_t + \frac{\partial g(N^*, P^*)}{\partial P} \rho_t$$

Let

$$a = \frac{\partial f(N^*, P^*)}{\partial N}, \; b = \frac{\partial f(N^*, P^*)}{\partial P}, \; c = \frac{\partial g(N^*, P^*)}{\partial N}$$

$$\text{and} \quad d = \frac{\partial g(N^*, P^*)}{\partial P}$$

and noticing that

$$f(N^*, P^*) = N^* \quad \text{and} \quad g(N^*, P^*) = P^*$$

we get

$$\eta_{t+1} = a\eta_t + b\rho_t$$

$$\rho_{t+1} = c\eta_t + d\rho_t$$

Solving for the eigenvalues of $\begin{bmatrix} a & b \\ c & d \end{bmatrix}$, say λ_1 and λ_2, we get

1. If $|\lambda_1| < 1$ and $|\lambda_2| < 1$, then we get stability because $\eta_t \rightarrow 0$ for η_0 sufficiently small
2. If $|\lambda_1| > 1$ or $|\lambda_2| > 1$, then we get instability.
3. Else no conclusion.

Graphical Methods. Graphical analyses of difference equation models are also possible (Fig. A.III.8). Equilibrium occurs when $N_{t+1} = N_t$, hence we may interpret the stability properties of equilibrium points that intersect a function, say $g(N_t)$, using the rules given above in step 5. This may be a very simple model or a complicated simulation model as analyzed by Holling et al. (1977), Peterman et al. (1979), and Gutierrez and Regev (1983).

Simple Models. Alternately, we may analyze the time course of the population. Assume that the initial populations is N_0 at time t_0 and during the time step it grows to N_{t+1} according to the dynamical rule $g(N_t)$. Recursively, we can continue the process as indicated in Figure A.III.8a, which is the discrete form of the exponential or Malthusian growth model ($N_{t+1} = (1 + r)N_t$). Note that the population continues to grow without bounds if $r > 0$. If we now include density dependent regulation using the discrete form of the logistic equation $\{N_{t+1} = N_t(1 + r(1 - N_t/K)] = g(N)N_t\}$, the population goes extinct with $r < 0$ (Fig. A.III.8b), but the population is stable at K when $0 < r < 1$ whether the stating density is below K (Fig. A.III.8b) or from above (Fig. A.III.8c). If, however, $2 > r > 1$, the function $g(N_t)$ declines after reaching K, the system approaches stability asymptotically (Fig. A.III.8d, see inset above). As r increases further, *bifurcation* occurs yielding oscillations of period two (Fig. A.III.8e) (see May and Oster 1976). If r is increased further, bifurcations arise yielding oscillations of period four. However, further increases in r produces chaotic behavior, but still further increases lead to regions of stability within the chaotic region, further bifurcations, and chaos.

The general topic of chaos is reviewed below and treated in a very readable manner by Gleick (1987) and Godfray and Grenfell (1993). The applicability of this theory to population ecology remains unclear, and is reviewed below.

Stability Properties Complicated Models. When the model is complicated, it may prove impossible to use analytical methods of analysis. The work of

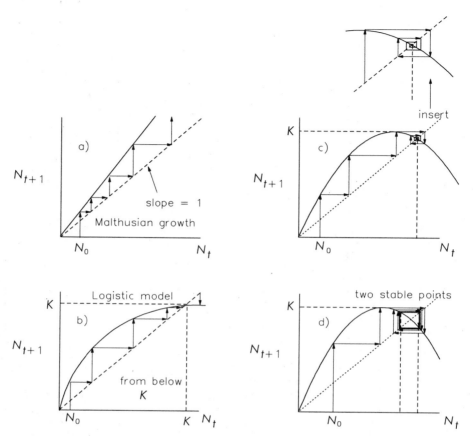

Figure A.III.8. *Graphical analyses of single species difference equation models: (a) Malthusian growth model, (b) the logistic model with $N_0 < K$, (c) an asymptotically stable model, and (d) an oscillatory stable model of period two (see the section on chaos, cf. May and Oster 1976).*

Holling and colleagues on the spruce budworm system in Northeastern Canada provides a convenient example for evaluating multitrophic interactions of biologically rich global models (Holling et al. 1977; Peterman et al. 1979). The details of the model are too complicated to review here, hence interested readers are referred to the original papers. Suffice it to say, the model evaluated below was a site simulation model for the balsam fir/spruce budworm/vertebrate natural enemy system that simulated different outcomes of spruce budworm populations across different levels of forest maturity.

Figure A.III.9a shows the results of numerous simulation runs where the ratio of budworm numbers (N) in one generation to those in the subsequent one [$R = N(t + 1)/N(t)$] across different levels of tree quality are plotted on spruce budworm density. As an aside, the troughs in the functions across tree quality are caused by the type III response of a vertebrate predators-to-prey

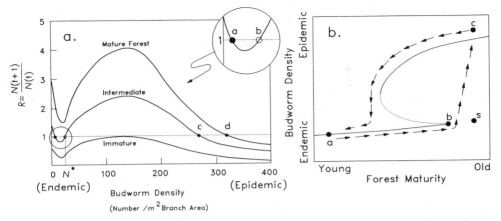

Figure A.III.9. *Analysis of the spruce budworm system: (a) Plots of simulation model results at various budworm densities across forest maturity classes, and (b) plots of equilibrium points (see inset in a.), where the points falling on the solid line are stable and those on the dashed line are unstable.*

density. The budworm populations would in theory be at equilibrium when the number of individuals in the following year were equal to those that produced them (i.e., $R = 1$). This is the criteria used to evaluate the stability properties of simple difference equation models of population interactions (see above), and since the simulation model is in reality the coding of a large difference equation model, the same methods apply. In this case, the functions depicted in Fig. A.III.9a intersect the horizontal line $R = 1$ at one or more levels of budworm populations (i.e., the abscissa) across different levels of forest maturity. The equilibrium points are stable only if $\uparrow N \Rightarrow \downarrow R$ or $\downarrow N \Rightarrow \uparrow R$ (i.e., an increase or decrease in N implies a decrease or increase respectively in R), while equilibria are unstable only if $\uparrow N \Rightarrow \uparrow R$ or $\downarrow N \Rightarrow \uparrow R$. The inset for one level of forest maturity in the figure shows that point a is stable but point b is unstable. Hence, for every function where three equilibria are generated, only two are stable.

These points are plotted on the phase plane of budworm density and forest maturity, where the stable points lie on the solid line and the unstable points lie on the dashed line (Fig. A.III.9b). Hence, if the budworm population begins at an endemic density and low forest maturity level indicated by point a, the budworm population equilibrium population level would increase as the forest maturity increased, but beyond point b the budworm population would quickly explode to epidemic levels indicated by point c, causing high mortality in older trees and a resulting rapid decline in forest maturity, and hence pest density. The model further suggests that control of the pest from level c to s would be pointless as the population would quickly resurge to point c. Equally, augmenting populations from low-endemic levels to near outbreak levels via mass immigration would result in the populations regressing quickly to endemic levels. This model captured the essence of the multiyear dynamics of this complicated system.

This method was bent to the analysis of strategies for energy allocation in the sylvan cotton–boll weevil system and its effect on the fitness and adaptedness of these two interacting species (see Chapter 10, Gutierrez and Regev 1983), where the problem was approached as the economics of resource acquisition and allocation.

CHAOS

What is the phenomenon of chaos? Lorenz (1963) who works in the field of meteorology, is credited with discovering the phenomena of chaos, and it has since been of increasing interest to theoretical ecologists. The phenomenon may arise in the solutions of simple difference equations used to model population dynamics (see above), but it may also arise in systems of differential equations (May and Oster 1976; Smale 1976). A feature of chaotic behavior is that the model exhibits sensitive dependence on initial conditions. In such models, a slight change in an initial condition may cause the model over time to exhibit very complex dynamics. The behavior of a chaotic system may appear noisy, but the results are outcomes of a deterministic process. For example, if one is interested in the behavior of a family of systems that differ only in a parameter value, the behavior of the chaotic system varies as the parameter varies, and can be described by the use of a bifurcation diagram (May and Oster 1976), where the position of the equilibria are plotted versus the parameter value. What results is a complex figure whose geometry shows self-similarity (i.e., the patterns repeat, Fig. A.III.10). For example, the portion of the figure near one bifurcation of, say, the three period cycle resembles the figure as a whole. Some argue that certain statistical features of the model output distinguish it from random noise. In these analyses, they use the Liapunov exponent, which measures the rate at which nearby trajectories diverge. The sign of this exponent is positive and finite in chaotic systems, whereas for uncorrelated random numbers it is theoretically infinite.

In contrast, Berryman and Millstein (1989) are critical of the idea that ecological systems are characterized by chaotic behavior. Berryman and Millstein (1989) argue that natural systems are unlikely to behave chaotically despite the fact they have the ingredients for such behavior [e.g., positive feedback loops (via reproduction) and delayed negative feedback loops which drive the system back into the positive feedback or growth region]. However, systems showing chaotic fluctuations would drop to low levels where the probability of extinction might be high. I would suppose that nature would select against reproductive strategies so as to avoid such uncontrolled fluctuations.

Evidence for Chaotic Behavior in Natural Systems

Many physical and chemical systems have been shown to behave in a manner that can be described by the mathematics of chaos. However, there seems to

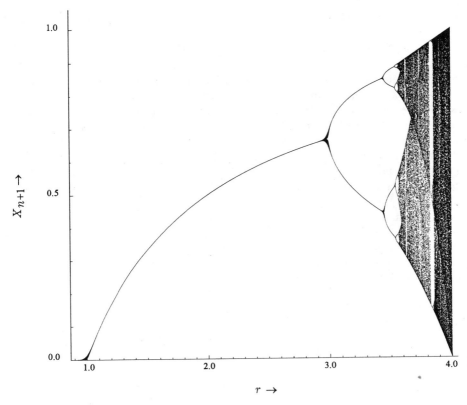

Figure A.III.10. *A bifurcation diagram of the discrete form of the logistic equation (May 1973; Turbo Pascal program by Stevens (1990).*

be no clear evidence that such behavior occurs in populations of natural systems. The detection of chaotic behavior in nature would depend on the analysis of long runs of non-noisy data, and such data are usually not available (Schaffer 1985). Schaffer claims that the fluctuations in the number of cases of measles during epidemics and the oscillations of lynx-hare densities contain evidence of chaotic behavior. This analysis remains a point of contention in the literature due likely to the inadequacy of the data. Schaffer and Knot (1986) expressed the view that the observed fluctuations in natural systems may be the result of underlying deterministic processes intrinsic to the system, as opposed to stochastic forcing. They further argue that if chaos is present in natural systems, the theoretical conclusions drawn about the system near equilibrium need to be reconsidered.

Hassell et al. (1991) suggest that a thorough understanding of the types of phenomena that are possible in nonlinear systems, and of the problems that such phenomenon pose for the design and interpretation of field studies is important if chaos can be shown to exist in natural systems. Hassell et al.

(1976) used the model

$$N(t + 1) = \lambda N(t)[1 + \alpha N(t)]^{-\beta}$$

to investigate chaotic behavior in single species populations using life table data from a number of insects. The parameter values were found to lie well within the monotonic damping range for the vast majority of the species they examined. For chaotic behavior to occur, the values of λ would have to be considerably larger than that observed. They concluded that chaotic behavior was unlikely to be observed, but as shown by them and Wang and Gutierrez (1980) for the simpler problem of point stability, this conclusion depends on the form of the model. Furthermore, the Hassell et al. (1976) analysis can be applied only to species with discrete generations in which density dependence is viewed as some process arising from within the population.

Recent work by Sugihara and May (1990) attempts to develop methods for detecting chaos in time series of noisy data. In the foreseeable future, the results of such analyses may not be amenable to testing in the field, as both the data and the models may prove inadequate.

An Overview of Some Optimization Methods Applicable to Applied Ecological Research*

The advantages and shortcomings of some optimization procedures applicable to applied ecology are discussed in this appendix only in an introductory way. Readers are referred to references cited for details. Familiarity with models used in crop pest management and a working knowledge of elementary differential calculus are assumed.

The problem of minimizing the function $f(x_1, x_2, \ldots, x_n)$ is identical to that of maximizing $-f$ (i.e., its negation). Hence, without loss of generality, consider the following general optimization problem

$$\max_{x_1, \ldots, x_n} f(x_1, x_2, \ldots, x_n) \qquad \text{[A.IV.1]}$$

subject to constraints that dictate the n biological interactions and restrictions (i.e., the parameter must not become unreasonable or violate assumptions).

$$g_i(x_1, x_2, \ldots, x_n) = 0 \qquad i = 1, 2, \ldots, p < n$$

$$g_i(x_1, x_2, \ldots, x_n) > 0 \qquad i = p + 1, \ldots, m$$

A general crop pest management problem may be formulated

$$f(x_1, x_2, \ldots, x_n) = B(x_1, x_2, \ldots, x_n) - C(x_1, x_2, \ldots, x_n) \qquad \text{[A.IV.2]}$$

where $B(\cdot)$ and $C(\cdot)$ are the gross revenue and the cost functions, respectively.

*Reprinted in modified form from Gutierrez and Way 1984.

Consider the following simple unrealistic example for the case $n = 1$. Suppose one wishes to know how much pesticide to apply at a specific time to maximize profit, with the assumptions that the number of pests killed is proportional (γ) to the amount of pesticides applied and the amount of damage is also proportional (α) to the number of pests present. Then the problem can be formulated such that

$$\text{revenues} = B(x) = p[Y_0 - \alpha q(1 - \gamma x)] \quad \text{and} \quad \text{costs} = C(x) = \beta x$$

$$[\text{A.IV.3}]$$

where Y_0 is the maximum yield under pest free conditions, q is the initial pest numbers; p is price per unit of yield, β is price per unit of pesticide (x), α and γ are constants of proportionality (see above).

The optimization problem [max $B(x) - C(x)$] can thus be written as

$$\max f(x) = p(Y_0 - \alpha q) + (\alpha q p \gamma - \beta)x$$

$$= pY_0 - \alpha q p + \alpha q p \gamma x - \beta x \qquad [\text{A.IV.4}]$$

subject to the constraints, and $\gamma x \leq 1$, $x \geq 0$. The terms pY_0 and $\alpha q p$ indicate the pest-free yield and the maximum yield loss for pest level q, respectively. The term $\alpha q p \gamma x$ denotes the benefit of pesticide use. The constraints indicate that one cannot kill more pests than are available, and the amount of pesticides to be applied is nonnegative ($x \geq 0$). The solution of [A.IV.4] is simple and is shown in Fig. A.IV.1. Note that the maximum depends on the value of the parameters and occurs either at 0 or at $1/\gamma$, which are the end points of the interval of x satisfying the constraints (i.e., *a corner solution*). This property is characteristic of *linear optimization problems*, for example, [A.IV.4]. In general, linear optimization problems take the form

$$\max_{y_1, \ldots, y_n} \sum_{j=1}^{} \gamma_j y_j \qquad [\text{A.IV.5}]$$

subject to $y_i \geq 0$, and

$$\sum_{j=1}^{} \gamma_{ij} y_j = b_i \quad \text{for} \quad i = 1, 2, \ldots, m \qquad [\text{A.IV.6}]$$

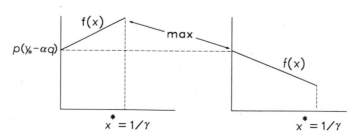

Figure A.IV.1. *The solution (x) for a simple linear optimization model.*

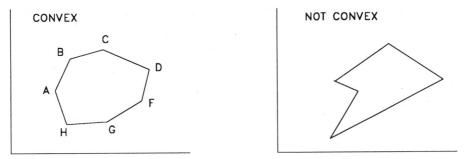

Figure A.IV.2. *A linear programming model solution: a convex solution and no solution.*

If there exists $(y_1^0, y_2^0, \ldots, y_n^0)$ satisfying [A.IV.5], then the region determined by these constraints is convex; for example that is, it has no holes, and its boundary has no indentations (i.e., not convex), as in Fig. A.IV.2. The solution of [A.IV.5] and [A.IV.6] occurs at an extreme point of this region (e.g., at corners or points A-H in the convex figure of Fig. A.IV.2). It can usually be obtained in a finite number of iterations using a fundamentally simple procedure (the *simplex method*, Dantzig 1951). If the linear programming problem is solved by enumeration, the number of required iterations is the number of extreme points of the region, as dictated by the constraints. For large problems these points can be numerous.

Consider a simple biological example to demonstrate the method. A predator has two choices of food A and B and its *objective* is to maximize the calories consumed. The *constraints* are

1. There are 100 calories available for search, and the cost of search is
   $\begin{cases} 2 \text{ calories per A} \\ 3 \text{ calories per B} \end{cases}$
2. There are 120 calories available for capture and the cost of capture is
   $\begin{cases} 4 \text{ calories per A} \\ 2 \text{ calories per B} \end{cases}$
3. The energy value of each prey item is $\begin{cases} 10 \text{ calories per A} \\ 20 \text{ calories per B} \end{cases}$

Next, we compute the constraint lines for search and for capture:

$$\textit{Search} \quad 2A + 3B \leq 100 \; (\text{———}) \quad \begin{cases} A = 0, \quad B = 33.3 \\ B = 0, \quad A = 50 \end{cases}$$

$$\textit{Capture} \; 4A + 2B \leq 120 \; (\text{-----}) \quad \begin{cases} A = 0, \quad B = 60 \\ B = 0, \quad A = 30 \end{cases}$$

and evaluate the corner solutions of the shaded polygram. In this case, the maximum occurs at B equal 33.3 suggesting that only it should be taken.

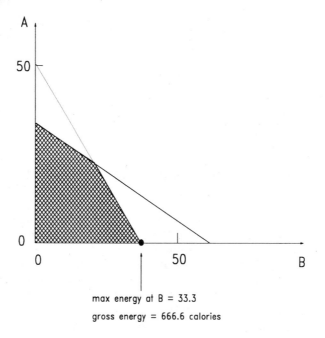

max energy at B = 33.3

gross energy = 666.6 calories

NONLINEAR PROBLEMS

The functions describing the biology of a crop ecosystem are rarely linear, hence in cases where the objective function or the constraints are nonlinear, methods of nonlinear and dynamic programming can be used. The formal development of the theory stems from differential calculus.

Consider a function $f(x)$ with one independent variable (Fig. A.IV.3). Then a point x^* is a maximum or minimum of $f'(x^*) = 0$ and $f''(x^*) < 0$ or > 0, respectively. For functions with more than one independent variable, similar conditions hold.[1]

Consider again a simple crop pest management problem similar to [A.IV.4]; only in this case the proportion of survived pests is a negative exponential of the amount of pesticides applied (see Fig. A.IV.4). This optimization problem becomes

$$\max_{x} g(x) = \max_{x} pY_0 - \alpha qpe^{-\gamma x} - \beta x \qquad [A.IV.7]$$

subject to the constraint $x \geq 0$.

[1]The maximum (minimum) of $f(x_1, x_2, \ldots, x_n)$ occurs at the point where the gradient of f, vector of partial derivatives, equals zero, and the matrix of second derivatives is negative (positive) definite.

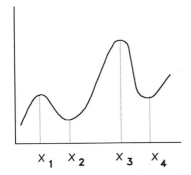

Figure A.IV.3. *The maximum and minimum of a function.*

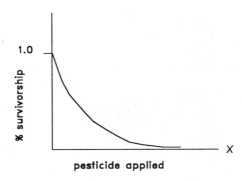

Figure A.IV.4. *Nonlinear pesticide survivorship model.*

To solve [A.IV.7], one first takes the derivative of the objective function

$$g'(x) = \gamma \alpha q p e^{-\gamma x} - \beta$$

and finds x^* such that $g'(x^*) = 0$. In this case,

$$x^* = \frac{1}{\gamma} \ln \left[\frac{\gamma \alpha q p}{\beta} \right]$$

Furthermore,

$$g''(x) = \gamma^2 \alpha q p e^{-\gamma x} < 0, \qquad \text{for all } x$$

hence the maximum of $g(x)$, in this case, is unique and occurs at x^*. However, [A.IV.7] is not completely solved because of the constraint that $x \geq 0$.

Case 1: $x^* > 0$. The solution of [A.IV.7] is then max $[g(0), g(x^*)]$, where $g(0) = pY_0 - \alpha q p$

and

$$g(x^*) = pY_0 - \beta / \gamma \left[1 + \ln \left(\frac{\gamma \alpha q p}{\beta} \right) \right]$$

Case 2: $x^* < 0$. This implies that $\gamma \alpha q p < \beta$, which means that for all $x \geq 0$, $g'(x) < 0$. Hence, $g(x)$ is a decreasing function of x, for $x \geq 0$, and the maximum thus occurs at $x = 0$.

In the above example we have illustrated how differential calculus is used to solve nonlinear optimization problems. However, this method has its limitations. To locate the set of points such that $g'(x) = 0$ can be difficult. In fact, a well-known theorem in algebra states that the zeros of a polynomial of degree 5 and higher cannot be solved in a finite number of algebraic computations.

For problems of great complexity such as a realistic crop–pest systems model, the techniques of differential calculus are usually inapplicable. In these cases the iterative methods of nonlinear optimization (i.e., programming) can be used. These iterative methods can be classified into two main categories: those that use differential calculus as a theoretical basis and those that do not. The general idea behind all iterative procedures is to start at a point, searching in some systematic way, locating another point that is closer to the optimum than the existing one and repeating the process. There are various ways of locating the next point, and this gives rise to a spectrum of nonlinear programming algorithms. Each method has its advantages and disadvantages. A disadvantage common to all iterative methods in nonlinear programming is that no systematic procedures exist to determine when a global optimum (the optimum across all of the search space) has been found, hence each must be considered a local optimum (i.e., one of many optima that may be found). Some optimization procedures for nonlinear problems are briefly described below, as a complete coverage is quite beyond our scope here.

METHODS BASED ON THE THEORY OF ELEMENTARY DIFFERENTIAL CALCULUS

These methods require information about the first and sometimes also the second derivatives of the objective function. Two of the best known classical optimization techniques of this category are the "steepest ascent" (Cauchy 1847; Goldstein 1962) and the Newton–Raphson method. The "steepest ascent," as the name indicates, searches in the steepest direction, which is that of the first derivative of the objective function. Its greatest shortcoming is that its rate of convergence is unpredictable near the optimum. The Newton–Raphson method, on the other hand, requires both the first and second derivatives, and it searches on the local quadratic surface. This procedure, contrary to "steepest ascent," is very efficient if the initial estimate is close to the optimum, otherwise the method may fail to converge. There are many extensions and variations of the above two procedures, among which is the method of Goldstein (1966).

While the property of rapid ultimate convergence, which we find in the Newton–Raphson method, is desirable, the evaluation of second derivatives is often costly. Hence, another group of algorithms was developed that used information of first derivatives to evaluate the second-degree terms. These methods are generally successful in practice, even though the reasons for success are as yet unknown. Some of these better known methods were those developed by Zoutendijk (1960), Fletcher and Reeves (1964), Powell (1964), and Davidon (1959); modifications are given by Fletcher and Powell (1963) and Stewart (1967). Stewart's procedure extends Davidon's variable metric methods. In cases where even the computation of first derivatives is laborious, Powell's and Stewart's algorithms can be used, as these procedures only require function evaluation (i.e., they can evaluate the entire algorithm simulation).

DIRECT SEARCH METHODS

Direct search methods have the following characteristics: (a) they use numerical and not analytical techniques, (b) they require only computations of function values, and (c) they can be applied to a wide class of optimization problems. These algorithms often compare favorably with the methods previously described. The classical method of changing one variable at a time falls into this category, as do those of Spendly et al. (1962) and Nelder and Mead (1965). The procedures discussed in the these papers are designed to solve optimization problems without constraints, but in any case there are ways to handle the constrained problems. Kelly (1960) and Rosen (1960, 1961) extended methods near programming to handle constraints that are linear, or nonlinear constraints that can be appropriately linearized. Another approach is to convert the constrained problem into an unconstrained one using Lagrange multipliers. An application of the constrained nonlinear programming methods to alfalfa-alfalfa weevil management was used in Regev et al. (1976). [Reviews of nonlinear optimization techniques by Powell (1970) and Polak (1973) may be helpful, as is the more complete biography on optimization in Leon (1965)].

DYNAMIC PROGRAMMING

Problems of the form of [A.IV.3] can sometimes be restructured so that the independent variables $X_1, X_2, \ldots, X_n$ form a sequence of interrelated decisions. The crop–pest management problem as described in the text falls naturally into this category, where X_1 is the optimal amount of pesticide (or control) to be applied during the ith time interval. For problems with this characteristic, a dynamic programming procedure can be used. This method, quite different from the methods discussed previously, was developed by Bellman and Dreyfus (1962). The advantages and disadvantages of this approach are discussed below. To transform the optimization problem to the dynamic programming framework, the following procedure is usually followed:

1. Determine the number of necessary state (e.g., pest levels) and decision variables (e.g., pesticide applications).
2. Define the optimal value function and its arguments (see example below).
3. Write the appropriate recurrence relation.
4. Compute the appropriate boundary conditions.

While the principle of dynamic programming is simple, good problem formulation requires experience and imagination (Dreyfus and Law 1977).

To illustrate the dynamic programming technique, consider an example similar to [A.IV.7], but in this case decisions of applying pesticides are made during two time periods. This requires that the dynamics of the problem in the first period be repeated in the second. The optimization problem can thus be written as

$$\max_{x_1, x_2} pY_0 \underbrace{-\alpha pqe^{-\gamma x_1}}_{\text{stage 1}} \underbrace{-\alpha pqe^{-\gamma x_1}e^{-\gamma x_2}}_{\text{stage 2}} - \beta(x_1 + x_2)$$

subject to $x_1 \geq 0$, $x_2 \geq 0$. The term $\alpha pqe^{-\gamma x_1}$ and $\alpha pqe^{-\gamma x_1}e^{-\gamma x_2}$ are the cost of uncontrolled pest damage in the first and second time periods, respectively. The exponential terms can be viewed as survivorship terms of pests to pesticide, hence in the second period the pests surviving the insecticide in the first period may have to survive another application in the second period. Note that the dynamics of the birth process are absent in the simple model. This problem can be reformulated so that it can be solved using dynamic programming. For simplicity, x_1 and x_2 take on values of 0 or 1 only, indicating a pesticide has been applied. Following the process indicated above, we have

1. The decision variables are x_1 and x_2. There are three stages; at stages 1 and 2 a decision to spray or not to spray is made, and stage 3 denotes the end of the season or time of harvest. The state variable is q, the current pest level.
2. The optimal value function is defined as $f_k(q) =$ the optimal return that can be obtained from stage k to stage $k + 1$, given that at stage k the pest level is q.
3. The recurrence relation can be written

$$f_k(q) = \max \begin{cases} \text{spray: } f_{k+1}(qe^{-\gamma}) - \alpha qpe^{-\gamma} - \beta & x = 1 \\ \text{not spray: } f_{k+1}(q) - \alpha pq & x = 0 \end{cases}$$

4. The boundary condition states that pY_0 is the maximum revenue

$$f_3(q) = f_3(qe^{-\gamma}) = f_3(qe^{-2\gamma}) = pY_0$$

The problem can be portrayed graphically as in Figure A.IV.5.
To numerically solve the dynamic programming problem, we make the following arbitrary assignments: $Y_0 = 10$ lb, $p = \$2$, $\alpha = 0.05$, $q = 50$, $\beta = \$4$ and $\gamma = 1$. Solving backwards, we have at stage 2

$$f_2(50) = \max \begin{cases} \text{spray (node 2} \rightarrow \text{5): } f_3(50e^{-1}) - 4 - (0.05)(2)(50)e^{-1} \\ \text{not spray (node 2} \rightarrow \text{4): } f_3(50) - (0.05)(2)(50) \end{cases}$$

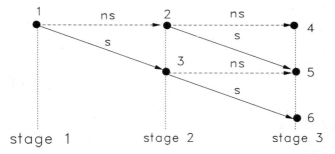

Figure A.IV.5. *A dynamic programming model where the decisions are to spray (s, ———) or not spray (ns, – – –) at stages 1 and 2 in order to maximize yields at stage 3.*

$$= \max \begin{cases} \text{spray} : 14.16 \\ \text{not spray} : 15.0 \end{cases} = 15.0 \text{ (not spray)}$$

$$f_2(50e^{-1}) = \max \begin{cases} \text{spray (node 3} \to \text{6): } f_3(50e^{-2}) - 4 - (0.05)(2)(50)e^{-2} \\ \text{not spray (node 3} \to \text{5): } f_3(50e^{-1}) - (0.05)(2)(50)e^{-1} \end{cases}$$

$$= \max \begin{cases} \text{spray: } 15.32 \\ \text{not spray: } 18.16 \end{cases} = 18.16 \text{ (not spray)}$$

and at stage 1

$$f_1(50) = \max \begin{cases} \text{spray (node 1} \to \text{3): } f_2(50e^{-1}) - 4 - (0.05)(2)(50)e^{-1} \\ \text{not spray (node 1} \to \text{2): } f_2(50) - (0.05)(2)(50) \end{cases}$$

$$= \max \begin{cases} \text{spray: } 12.32 \\ \text{not spray: } 10.0 \end{cases} = 12.32 \text{ (spray)}$$

Hence, the optimum solution is to spray at stage 1 (i.e., $x_1 = 1$) and to not spray at stage 2 (i.e., $x_2 = 0$), and the return is 12.32. As a check, the brute-force enumeration gives the following results: $x_1 = 0$, $x_2 = 0$, return $= 10.0$; $x_1 = 1$, $x_2 = 0$, return $= 12.32$; $x_1 = 0$, $x_2 = 1$, return $= 9.16$; and $x_1 = 1$, $x_2 = 1$, return $= 9.48$. The results of the four different paths are shown in Figure A.IV.6 showing that the best option (Fig. A.IV.6C) is to spray in the first period but not in the second.

The advantage of dynamic programming over brute-force enumerations is not evident from the above simple example. The power of technique can be shown by considering an equipment replacement problem where a decision of whether to replace or not to replace is made monthly, and the objective is to minimize cost over 20 years. If we have available a digital computer that takes 10^{-5} s to perform an addition or comparison, then the computer solution of

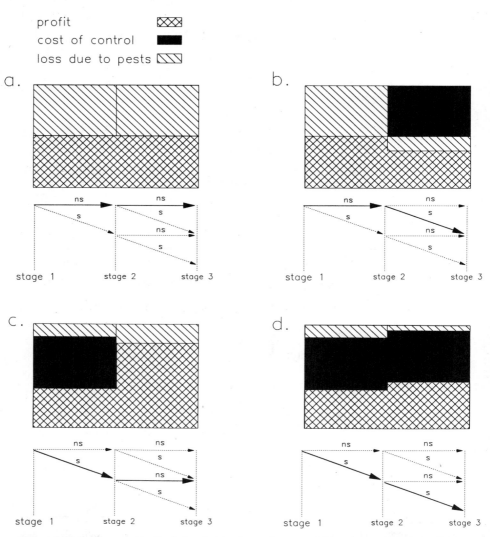

Figure A.IV.6. *The graphical solution (→) to the various possible spray combinations of the dynamic programming model in Figure A.IV.5. Note that the whole rectangle equals pY_0.*

the dynamic programming problem takes 1.5 sec. However, if we were to use brute-force enumeration, the same computer would require 10^{62} year (Dreyfus and Law 1977). The machine replacement dynamic programming problem requires only one state variable and was efficiently solvable, but as the number of state variables increases relative to the number of possible decisions, the advantage of dynamic programming over brute-force enumeration disappears.

Consider the following general crop–pest management problem

$$\max_{x_1, x_2, \ldots, x_{20}} \quad B(., x_1, x_2, \ldots, x_{20}) - C(x_1, x_2, \ldots, x_{20})$$

where $B(\cdot)$ and $C(\cdot)$ are as defined in the text and the decision to apply or not to apply pesticides at the ith time interval is indicated by x_i. Furthermore, suppose that in the dynamic programming formulation of the problem, it is necessary to retain as state variables 10 age classes each of leaves and fruit, and each age class has 10 possible values. For this problem, then, the brute-force enumeration requires 2^{20} at 1.05×10^6 calculations of $(B - C)(\cdot)$, and the dynamic programming solution takes $2 \times 20 \times 10^{20}$ or 4×10^{21} function value computations. Hence, the applicability of the technique of dynamic programming to crop–pest management depends among other things on the separability of the crop ecosystem simulation model from the optimization problem. In cases where the amount of damage can be determined from the number of pests alone, dynamic programming can probably be used to solve efficiently the optimization problem. If, however, the optimal value function depends also on the state of the plant, the dynamic programming procedure would most likely be impractical. Hence, the major limitation of dynamic programming to complex problems is the inability to handle even a moderate number of state variables. In cases where the dynamic programming procedure is not feasible, methods of nonlinear programming mentioned in previous sections might be more suitable.

This is but a brief introduction to some existing optimization methods that might be applicable to crop or animal pest management. Rapid progress is being made on several fronts not covered here.

References

Abrams, P. 1994. The fallacies of 'ratio-dependence.' Ecology 75: 1842–1850.

Adkisson, P. L., R. A. Bell, and S. G. Wellso. 1963. Environmental factors controlling the induction of diapause in the pink bollworm *Pectinophora gossypiella* (Sanders) J. Insect Physiol. 9: 299–310.

Ågrens, G. I. and E. Bosatta. 1990. Theory and model or art and technology in ecology. Ecol. Modelling 50: 213–220.

Akçakaya, H. R., L. R. Ginsburg, L. R. Slice, and S. L. Slobodkin. 1988. The theory of population dynamics: II physiological delays. Bull. Math. Biol. 50: 503–515.

Allen, J. C. 1976. A modified sine wave method for calculating degree days. Environ. Entomol. 5: 388–396.

Andrewartha, H. G. and L. C. Birch. 1954. *The Distribution and Abundance of Animals.* Chicago: University of Chicago Press.

Anderson, R. M. and R. M. May. 1980. Infectious disease and population cycles of forest insects. Science 210: 658–661.

Arditi, R. and A. A. Berryman. 1991. The biological control paradox. Trends in Ecology and Evol. 6:32.

Arditi, R. and L. R. Ginzburg. 1989. Coupling in predator-prey dynamics: ratio dependence. J. Theor. Biol. 139: 311–326.

Arditi, R. and L. Ginsberg. 1992. Scale invariance is a reasonable approximation in predation models—reply to Ruxton and Gurney. OIKOS 65: 336–337.

Arditi, R., N. Perrin, and H. Saiah. 1991. Functional response and heterogeneities: an experimental test with cladocerans. OIKOS 60: 69–75.

Arditi, R. and J. Michaelski. 1995. Nonlinear food web models and their response to increased basal productivity. In: G. A. Polis and K. O. Winemiller, (eds.). *Food webs: integration of patterns and dynamics.* Chapman and Hall, London, England.

Arditi, R. and H. Saïah. 1992. Empirical evidence for the role of heterogeneity in ratio dependent consumption. Ecology 73: 1544–1551.

Arrhenius, S. A. 1889. Ueber die reaktionsgeschwindigkeit bei der inversion von rohrzucker durch sauren. Ztschr. Physikal. Chem. 4: 226–248.

Baumgärtner, J. U., M. Beri, and V. Delucchi. 1987. Growth and development of immature life stages of *Propylaea 14-punctata* L. and *Coccinella 7-punctata* L. [Col.: Coccinellidae] simulated by the metabolic pool model. Entomophaga 32: 415–423.

Baumgärtner, J. U. and O. Bonato. 1991. Analyse et gestion des agroecosystémes: Utilisation d'une approche demographique. Rech. Agron. Suisse 31: 177–191.

Baumgärtner, J., V. Delucchi, R. von Arx, and D. Rubli. 1986b. Whitefly (*Bemisia tabaci* Genn., Stern.: Aleyrodidae) infestation patterns as influenced by cotton, weather and *Heliothis*: Hypotheses testing by using simulation models. Agric. Ecosyst. Environ. 17: 49–59.

Baumgärtner, J. U., M. Genini, B. Graf and A. P. Gutierrez. 1986. Generalizing a population model for simulating Golden delicious apple tree growth and development. *Acta Hortic.*, 184: 111–122.

Baumgärtner, J. and A. P. Gutierrez. 1989. Simulation techniques applied to crops and pest models. In (R. Cavalloro, and V. Delucchi (eds.): *Parasitis 88. Proceedings of a scientific congress*, Barcelona, 25–28 October, Boletin de Sanidad Vegetal, Fuera de Serie, vol. 17, pp. 175–214.

Baumgärtner, J. U., A. P. Gutierrez, and C. S. Summers. 1981. The influence of aphid prey consumption on searching behavior, weight increase, developmental time, and mortality of *Chrysopa carnea* Stephens and *Hippodamia convergens* G-M. Can. Entomol. 113: 1015–1024.

Baumgärtner, J. and M. Severini. 1987. Microclimate and arthropod phenologies: the leaf miner *Phyllonorycter blancardella* F. (Lep.) as an example. In F. Prodi, F. Rossi, and G. Cristoferi, (eds.): *Agrometerology*. Bologna: Editrice Compositori, 498 pp.

Baumgärtner, J., M. Severini, and M. Ricci. 1990. The mortality of overwintering *Phyllonorycter blancardella* (Lep., Gracillariidae) pupae simulated as a loss in a time-varying distributed delay model. Mitt. Schweiz. Entomol. Ges. 63: 439–450.

Beddington, J. R., M. P. Hassell, and J. H. Lawton. 1976. The components of arthropod predation. II. The predator rate of increase. J. Anim. Ecol. 45: 165–185.

Bellman, R. E. and S. E. Dreyfus. 1992. Applied Dynamic Programming. Princeton, Princeton University Press.

Berryman, A. A. 1981. *Population Systems: A General Introduction*. New York: Plenum.

Berryman, A. A. 1991. Vague notions of density-dependence. OIKOS 62: 252–254.

Berryman, A. A. 1992. The origins and evolution of predator–prey theory. Ecology 73: 1530–1573.

Berryman, A. A., J. Michaelski, A. P. Gutierrez, and R. Arditi. 1995. Logistic theory of food web dynamics. Ecology 76: 336–343.

Berryman, A. A. and J. A. Millstein. 1989. Are ecological systems chaotic and if not why not? Trends in Ecology and Evol. 4: 26–28.

Berryman, A. A. and N. C. Stenseth. 1984. Behavioral catastrophes in biological systems. Behav. Sci. 29: 127–137.

von Bertalanffy, L. 1962. Modern Theories of Development: An Introduction to Theoretical Biology. Harper and Brothers, New York: Torchbooks (originally published in 1933 by Oxford University Press, translated by J. H. Woodger).

Bianchi, G., J. Baumgärtner, V. Delucchi, and N. Rahalivavololona. 1990. Modéle de population pour la dynamique de *Maliarpha separatella* Rag. (Pyralidae, Phycitinae) dans les rizières malgaches du lac Alaotra. J. Appl. Entomol. 110: 384–397.

Bianchi, G., J. U. Baumgärtner, V. Delucchi, N. Rahalivavololona, S. Skillman, and P. Zahner. 1989. Sampling egg batches of *Maliarpha separatella* RAG (Lep., Pyralidae) in Madagascan rice fields. Trop. Pest Manag. 35: 420–424.

Burke, H. R. 1976. Bionomics of anthonomine weevils. Ann. Rev. Entomol. 21: 283–303.

Byerly, K. F., A. P. Gutierrez, R. E. Jones, and R. F. Luck. 1978. A comparison of sampling methods for some arthropod populations in cotton. Hilgardia 46: 257–282.

Cain, A. J. and P. M. Sheppard. 1950. Selection in the polymorphic land snail *Cepaea nemoralis*. Heredity, London 4: 275–294.

Campbell, A., B. D. Frazer, N. Gilbert, A. P. Gutierrez, and M. Mackauer. 1974. The temperature requirements of some aphids and their parasites. J. Appl. Ecol. 11: 431–438.

Candolle, A. P. de. 1855. *Geographique Botanique*. Paris: Raisonee.

Carey, J. R. 1982. Demography and population dynamics of the mediterranean fruit fly. Ecol. Modelling 16: 125–150.

Carey, J. R. 1993. *Applied Demography for Biologists*. New York: Oxford University Press, 206 p.

Carson, R. 1962. *Silent Spring*. Boston: Houghton Mifflin Company, 369 pp.

Carter, N., D. P. Aikman, and A. F. G. Dixon. 1973. An appraisal of the Hughes' time specific life table analysis for determining aphid reproduction and mortality rates. J. Anim. Ecol. 47: 677–688.

Casas, J. 1989. Foraging behavior of a leafminer parasitoid in the field. Ecol. Entomol. 14: 257–265.

Casti, J. and A. Karlqvist. 1989. *Newton to Aristotle—Toward a theory of models for living systems*. Boston: Birhauser, 284 pp.

Cauchy, A. L. 1847, C. R. Read Sci., Paris 25: 536–538.

Cave, R. D. and A. P. Gutierrez. *Lygus hesperus* field life table studies in cotton and alfalfa (Heteroptera: Miridae). Can. Entomol. 115: 649–654.

Cerutti, F., J. Baumgärtner, and V. Delucchi. 1991. The dynamics of grape leafhopper *Empoasca vitis* Gothe populations in Southern Switzerland and the implications for habitat management. Biocontrol Sci. Tech. 1: 177–194.

Chapin, F. S., E. Schulze, and H. A. Mooney. 1990. The ecology and economics of storage in plants. Ann. Rev. Ecol. Syst. 21: 399–422.

Chapman, R. F. 1982. *The Insects: Structure and Function*. Third addition. Cambridge: Harvard University Press, 919 pp.

Cherlet, M. and A. DiGregorio. 1991. Calibration and integrated modelling of remote sensing data for desert locus habitat monitoring. FAO remote sensing center report ECLO/INT/004/BEL and GCP/INT/BEL Rome.

Chesson, P. L. 1981. Models of spatially distributed populations: the effect of within-patch variability. J. Theor. Popul. Biol. 19: 288–325.

Chesson, J. 1983. The estimation and analysis of preference and its relationship to foraging models. Ecology 64: 1297–1304.

Chi, H. 1988. Life-table analysis incorporating both sexes and variable development rates among individuals. Environ. Entomol. 17: 26–34.

Chi, H. and W. M. Getz. 1988. Mass rearing and harvesting based on an age-stage, two-sex life table: A potato tuberworm (Lepidoptera: Gelechiidae) case study. Environ. Entomol. 17: 18–25.

Cock, M. J. W. 1978. The assessment of preference. J. Anim. Ecol. 47: 805–816.

Conway, G. R. (ed.) 1984. *Pest and Pathogen Control: Strategic, Tactical, and Policy Models*. New York: Wiley, p. 159–183.

Costanza, R. 1993. Developing ecological research that is relevant for achieving sustainability. Ecological Appl. 3: 379–381.

Courant, R. and D. Hilbert. 1962. *Methods of Mathematical Physics*, Vol. II. New York, London, and Sydney: Interscience.

Crawley, M. J. 1992. *Natural Enemies: Biology of Predators, Parasites and Diseases*. London: Blackwell Scientific Publications, 576 pp.

Cracknell, A. P. and L. Hayes, 1991. *Introduction to Remote Sensing*. New York: Taylor & Francis, 293 pp.

Crichton, M. 1990. *Jurassic Park*. Ballantine Books, New York.

Curry, G. L. and R. M. Feldman. 1987. *Mathematical Foundations of Population Dynamics*. College Station, TX TEES Monograph Series No. 3.

DaSilva, P. G., K. S. Hagen, and A. P. Gutierrez. 1992. Functional response of *Curinus coleruleus* (Col.: Coccinellidae) to *Heteropsylla cubana* (Hom.: Psyllidae) on artificial and natural substrates. Entomophaga 37: 555–564.

Danzig, G. B. 1951. *Maximization of Linear Functions of Variables Subject to Linear Inequalities, Activity Analyses of Production and Allocation*. New York: Wiley.

Davidon, W. C. 1959. *Variable Metric Methods for Minimization*. A.E.C. Res. Dev. Rep. ANL 5990.

DeAngelis, D. L., R. A. Goldstein, and R. V. O'Neill. 1975. A model for trophic interactions. Ecology 56: 881–892.

DeAngelis, D. L. and L. J. Gross (eds.). 1992. *Individual Based Models and Approaches in Ecology, Populations, Communities and Ecosystems*. New York: Chapman and Hall.

Deevey, E. S. 1947. Life tables for natural populations of animals. Quart. Rev. Biol. 22: 283–314.

Dixon, A. F. G. 1987. Parthenogenetic reproduction and the rate of increase in aphids. In *Aphids: Their Biology, Natural Enemies and Control* (vol. A). Amsterdam: Elsevier, pp. 269–287.

DosSantos, W. J., A. P. Gutierrez, and M. A. Pizzamiglio. 1989. Evaluating the economic damage caused by the cotton stem borer *Eutinobothrus brasiliensis* (Hambleton 1937) in cotton in southern Brasil. Pesqui. Agropecu. Bras. 24: 297–305.

Dreyfus, S. E. and A. M. Law. 1977. *The Art and Theory of Dynamic Programming.* New York: Academic, 284 pp.

Dublin, L. I. and A. J. Lotka. 1925. On the true rate of natural increase. J. Am. Stat. Assoc. 20: 305–339.

Efron, B. 1982. The Jack-knife, the Bootstrap and other related sampling plans. CBMS-NSF Regional Conference series in Applied Mathematics. Society of Industrial Applied Mathematics, Philadelphia, PA.

Elton, C. S. 1939. *Animal Ecology.* New York: Macmillan.

Elton, C. S. 1942. *Voles, mice and lemmings: problems in population dynamics.* Oxford: Oxford University Press.

Elton, R. A. and J. J. D. Greenwood. 1970. Exploring apostatic selection. Heredity, London 25: 629–633.

Ehrlich, P. 1993. Science and the management of natural resources. Ecol. Appl. 3: 358–360.

Falcon, L. A., R. van den Bosch, J. Gallegher, and A. Davidson. 1971. Investigation of the pest status of *Lygus hesperus* in cotton in Central California. J. Econ. Entomol. 64: 56–61.

Fisher, R. A. 1930. *The Genetical Theory of Natural Selection.* Oxford: Clarendon.

Fischlin, A. and Baltensweiler, W. 1979. Systems analysis of the larch bud moth system. Part 1: the larch–larch bud moth relationship. In V. Delucchi and W. Baltensweiler (eds.): *Dispersal of forest insects.* Zuoz, Switzerland: IUFRO, pp. 273–289.

Fletcher, R. and M. J. D. Powell, 1963. A rapidly convergent descent method for minimization. Computer J. 6: 163–168.

Fletcher, R. and C. M. Reeves. 1964. Function minimization of conjugate gradients. Computer J. 7: 149–154.

Flint, M. L. 1981. Climatic ecotypes of *Trioxys complanatus*, a parasite of the spotted alfalfa aphid. Environ. Entomol. 9: 501–507.

Foerster, von H, 1959. Some remarks on changing populations. In F. Stahlman, Jr. (ed.): *The Kinetics of Cellular Proliferation.* New York: Grune and Stratton, pp. 382–407.

Force, D. C. and P. S. Messenger. 1964. Fecundity, reproductive rates, and innate capacity for increase of three parasites of *Therioaphis maculata* (Buchton). Ecology: 45: 706–715.

Fracker, S. B. and H. A. Brischle. 1944. Measuring the local distribution of *Ribes.* Ecology 25: 283–303.

Frazer, B. D. and N. Gilbert. 1976. Coccinellids and aphids: a quantitative study of the impact of adult ladybirds (Coleoptera: Coccinellidae) preying on field populations of pea aphids (Homoptera: Aphididae). J. Entomol. Soc. Br. Columbia 73: 33–56.

Frazer, D. A. S. and I. Guttman. 1956. Tolerance regions. Ann. Math. Statist. 27: 162–79.

Free, C. A., J. R. Beddington, and J. H. Lawton. 1977. On the inadequacy of simple models of mutual interference for parasitism and predation. J. Anim. Ecol. 46: 543–554.

Freund, J. E. 1962. *Mathematical Statistics*. Englewood Cliffs NJ: Prentice Hall, 390 pp.

Freund, J. E. and R. E. Walpole 1987. *Mathematical Statistics*, 4th ed. Englewood Cliffs NJ: Prentice Hall, 608 pp.

Fitzpatrick, E. A. and H. A. Nix. 1970. The climatic factor in Australian grasslands ecology. In R. M. Moore (ed.): *Australian Grasslands* Australian National University Press, pp. 3–26.

Fuentes, E. R. 1993. Scientific research and sustainable development. Ecol. Appl. 3: 376–377.

Garcia, R., L. E. Caltagirone, and A. P. Gutierrez. 1988. Comments on the redefinition of biological control. BioScience 38: 692–694.

Gatto, M., C. Matessi, and L. B. Slobodkin. 1989. Physiological profiles and demographic rates in relation to food quantity and predictability: an optimization approach. Evol. Ecol. 3: 1–30.

Gause, G. F. 1934. *The struggle for existence*. Baltimore: Williams and Wilkins.

Getz, W. M. 1984. Population dynamics: a per capita resource approach. J. Theor. Biol. 108: 623–643.

Getz, W. M. 1991. A unified approach to multispecies modelling. Nat. Res. Modeling 5: 393–421.

Getz, W. M. and A. P. Gutierrez, A perspective on systems analysis in crop production and insect pest management. Ann. Rev. Entomol. 27: 447–66.

Getz, W. M. and R. C. Haight. 1989. *Population Harvesting: demographic models of fish, forest and animal reosurces*. Princeton NJ: Princeton University Press, 390 pp.

Gilbert, F. (ed): 1990. *Insect life cycles: genetics, evolution and co-evolution*. Berlin: Springer-Verlag, 258 pp.

Gilbert, N. 1981. Comparison of predation rates. Can. Entomol. 113: 1047–1048.

Gilbert, N. 1984. What they didn't tell you about limit cycles. Oecologia 65: 112–113.

Gilbert, N. and A. P. Gutierrez. 1973. A plant-aphid-parasite relationship. J. Anim. Ecol. 42: 323–340.

Gilbert, N., A. P. Gutierrez, R. E. Jones and B. D. Frazer. 1976. *Ecological Relationships*. New York: Freeman.

Ginsberg, L. R. and H. R. Akçakaya. 1992. Consequences of ratio-dependent predation for steady state properties of ecosystems. Ecology 73: 1536–1543.

Gleick, J. 1987. *Chaos: making a new science*. New York: Viking Penguin Inc., 352 pp.

Godfray, H. C. J. and B. T. Grenfell. 1993. The continuing quest for chaos. Trends in Ecology and Evol. 8: 43–44.

Godfray, H. C. J. and J. K. Waage. 1991. Predictive modelling in biological control: the mango mealybug (*Rastococcus invadens*) and its parasitoids. J. Appl. Ecol. 23: 434–453.

Goldstein, M. 1966. Graph Theoretic Codes: Research Project. UCB, Department of Electrical Engineering and Computer Sciences.

Goldstein, N. E. 1962. Numerical Filtering of Potential Field Signals as Applied to Geophysical Exploration. Thesis (M.S.) University of California, Berkeley.

Gordon, H. T. 1984. Growth and development in insects. In C. B. Huffaker and R. L. Rabb, (eds.): *Ecological Entomology*, New York: Wiley, pp. 53–72.

Gossard, T., C. G. Summers, and A. P. Gutierrez. 1976. Factors affecting predation by *Hippodamia convergens* Guerin-Meneville (Coleoptera: Coccinellidae). Division of Biological Control Progress Report, University of California, Berkeley.

Goudriaan, J. 1973. Dispersion in simulation models of population growth and salt movement in the soil. Neth. J. Agric. Sci. 21: 269–281.

Graf, B., A. P. Gutierrez, O. Rakotobe, P. Zahner, and V. Delucchi. 1990a. A simulation model for the dynamics of rice growth and development. II. Competition with weeds for nitrogen and light. Agric. Syst. 32: 367–392.

Graf, B., O. Rakotobe, P. Zahner, V. Delucchi, and A. P. Gutierrez. 1990b. A simulation model for the dynamics of rice growth and development Part I—The carbon balance. Agric. Syst. 32: 341–366.

Gurney, W. S. C., S. P. Blythe, and R. M. Nisbet. 1980. Nicholson's blowflies revisited. Nature (London) 287: 17–21.

Gurney, W. S. C., R. M. Nisbet, and J. H. Lawton. 1983. The systematic formulation of tractable single-species population models incorporating age structure. J. Anim. Ecol. 52: 479–495.

Gutierrez, A. P. 1992. The physiological basis of ratio dependent theory. Ecology 73: 1552–63.

Gutierrez, A. P. 1995. Integrated pest management in cotton. In D. Dent (ed.): *Integrated Pest Management*. London: Chapman and Hall (in press).

Gutierrez, A. P. and J. U. Baumgärtner. 1984a. Multitrophic level models of predator-prey–energetics: I. Age specific energetics models-pea aphid *Acyrthosiphon pisum* (Harris) (Homoptera: Aphididae) as an example. Can. Entomol. 116: 924–932.

Gutierrez, A. P. and J. U. Baumgärtner, 1984b. Multitrophic level models of predator-prey–energetics: II. A realistic model of plant–herbivore–parasitoid–predator interactions. Can. Entomol. 116: 933–949.

Gutierrez, A. P., J. U. Baumgärtner, and K. S. Hagen. 1981a. A conceptual model for growth, development and reproduction in the ladybird beetle *Hippodamia convergens* G.M. (Coccinellidae: Coleoptera). Can. Entomol. 113: 21–33.

Gutierrez, A. P., J. U. Baumgärtner, and C. G. Summers. 1984a. Multitrophic level models of predator–prey energetics: III. A case study of an alfalfa ecosystem. Can. Entomol. 116, 950–963.

Gutierrez, A. P., G. D. Butler Jr., and C. K. Ellis. 1981. Pink bollworm diapause induction and termination in relation to fluctuating temperatures and decreasing photoperiod. Environ. Entomol. 10: 936–42.

Gutierrez, A. P., G. D. Butler Jr., Y. Wang, and D. Westphal 1977a. The interaction of pink bollworm (Lepidoptera: Gelichiidae), cotton, and weather: a detailed model. Can. Entomol. 109: 1457–1468.

Gutierrez, A. P., J. B. Christensen, C. M. Merritt, W. B. Loew, C. G. Summers, and W. R. Cothran. 1976. Alfalfa and the Egyptian alfalfa weevil. Can. Entomol. 108: 635–648.

Gutierrez, A. P. and G. L. Curry. 1989. Conceptual framework for studying crop–pest systems. In R. E. Frisbie, K. M. El-Zik, and L. T. Wilson (eds.): *Integrated Pest Management Systems and Cotton Production*. New York: Wiley, pp. 37–64.

Gutierrez, A. P. and R. Daxl. 1984. Economic threshold for cotton pests in Nicaragua: ecological and evolutionary perspectives. In G. R. Conway (ed.): *Pests and pathogens control: strategic, tactical and policy models*. New York: Wiley, pp. 184–205.

Gutierrez, A. P., R. Daxl, G. Leon Quant and L. A. Falcon. 1981b. Estimating economic thresholds for bollworm, *Heliothis zea* Boddie and boll weevil, *Anthonomus grandis* Boh., damage in Nicaraguan cotton, *Gossypium hisutum* L., Environ. Entomol. 10: 873–879.

Gutierrez, A. P., W. J. Dos Santos, M. A. Pizzamiglio, A. M. Villacorta, C. K. Ellis, C. A. P. Fernandes, and I. Tutida. 1991a. Modelling the interaction of cotton and the cotton boll weevil. II. Boll weevil (*Anthonomus grandis*) in Brazil. J. Appl. Ecol. 28: 398–418.

Gutierrez, A. P., W. J. Dos Santos, A. Villacorta, M. A. Pizzamiglio, C. K. Ellis, L. H. Carvalho, and N. D. Stone. 1991b. Modelling the interaction of cotton and the cotton boll weevil. I. A comparison of growth and development of cotton varieties. J. Appl. Ecol. 28: 371–397.

Gutierrez, A. P., L. A. Falcon, W. Loew, P. A. Leipiz, and R. van den Bosch. 1975. An analysis of cotton production in California: a model for Acala cotton and the effects of defoliators on yield. Environ. Entomol. 4: 125–36.

Gutierrez, A. P., K. S. Hagen, and C. K. Ellis 1990. Evaluating the impact of natural enemies: a multitrophic perspective. In M. Mackauer, L. E. Ehler, and J. Roland (eds.): *Critical Issues in Biological Control*. Andover, Hants, UK: Intercept, pp. 81–109.

Gutierrez, A. P., D. E. Havenstein, H. A. Nix, and P. A. Moore. 1974a. The ecology of *Aphis craccivora* Koch and subterranean clover stunt virus in Southeast Australia. II. A model of cowpea aphid populations in temperate pastures. J. Appl. Ecol. 11: 1–20.

Gutierrez, A. P., D. E. Havenstein, H. A. Nix, and P. A. Moore. 1974b. The ecology of *Aphis craccivora* Koch and subterranean clover stunt virus. III. A regional perspective of the phenology and migration of the cowpea aphid. J. Appl. Ecol. 11: 21–35.

Gutierrez, A. P., T. F. Leigh, Y. Wang, and R. D. Cave. 1977b. An analysis of cotton production in California: *Lygus hesperus* (Heteroptera: Miridae) injury—an evaluation. Can. Entomol. 109: 1375–1386.

Gutierrez, A. P., E. Mariot, J. R. Hakim Cure, and A. Villacorta. 1993. A model for the growth and development of three varieties of common bean (*Phaseolus vulgaris* L.): Factors affecting yield and quality. Agric. Syst. 44: 35–63.

Gutierrez, A. P., N. J. Mills, S. J. Schreiber, and C. K. Ellis 1994. A physiologically based tritrophic perspective on bottom up–top down regulation of populations. Ecology 75: 2227–2242.

Gutierrez, A. P., D. J. Morgan, and D. E. Havenstein. 1971. The ecology of *Aphis craccivora* Koch and subterranean clover stunt virus. I. The phenology of aphid populations and the epidemiology of virus in pastures in Southeast Aust. J. Appl. Ecol. 8: 699–721.

Gutierrez, A. P., P. Neuenschwander, F. Schulthess, B. Wermelinger, H. R. Herren, J. U. Baumgärtner, and C. K. Ellis, 1988b. Analysis of the biological control of cassava pests in West Africa, II. The interaction of cassava and cassava mealybug. J. Appl. Ecol. 25: 921–940.

Gutierrez, A. P., M. A. Pizzamiglio, W. J. Dos Santos, R. Tennyson and A. M. Villacorta. 1984b. A general distributed delay time varying life table plant population model: cotton (*Gossypium hirsutum* L.) growth and development as an example. Ecol. Modelling 26: 231–249.

Gutierrez, A. P. and U. Regev. 1983. The Economics of fitness and adaptedness: the interaction of sylvan cotton (*Gossypium hirsutum* L.) and the boll weevil (*Anthonomus grandis* Boh.) An example. Acta Oecolog. 4: 271–287.

Gutierrez, A. P., F. Schulthess, L. T. Wilson, A. M. Villacorta, C. K. Ellis, and J. U. Baumgärtner. 1987. Energy acquisition and allocation in plants and insects: a hypothesis for the possible role of hormones in insect feeding patterns. Can. Entomol. 119: 109–129.

Gutierrez, A. P., J. J. M. van Alphen, and P. Neuenschwander. 1993. Factors affecting the establishment of natural enemies: biological control of the cassava mealybug in West Africa by introduced parasitoids: a ratio dependent supply–demand driven model. J. Appl. Ecol. 30: 706–721.

Gutierrez, A. P. and R. van den Bosch. 1971. Studies on host selection and host specificity of the aphid hyperparasite *Charips victrix* (Hymenoptera: Cynipidae). 1. Review of hyperparasitism and the field ecology of *Charips victrix*. Ann. Entomol. Soc. Am. 63: 1345–1354.

Gutierrez, A. P. and Y. H. Wang. 1977. Applied population ecology: models for crop production and pest management. In G. A. Norton and C. S. Holling (eds.): *Pest Management*, Pergamon, Oxford: International Institute for Applied Systems Analysis Proceedings Series.

Gutierrez, A. P. and Y. H. Wang. 1984. Models for managing the economic impact of pest populations in agricultural crops. In C. B. Huffaker and R. L. Rabb (eds.): *Ecological Entomology*, New York: Wiley, pp. 729–763.

Gutierrez, A. P., B. Wermelinger, F. Schulthess, C. K. Ellis, J. U. Baumgärtner, and S. J. Yaninek. 1988a. Analysis of the biological control of cassava pests in West Africa, I. Simulation of carbon, nitrogen and water dynamics in cassava. J. Appl. Ecol. 25: 901–920.

Gutierrez, A. P., D. W. Williams, and H. Kido. 1985. A model of grape growth and development: the mathematical structure and biological considerations. Crop Sci. 25: 721–728.

Gutierrez, A. P. and J. S. Yaninek. 1983. Responses to weather of eight aphid species commonly found in pastures in southeastern Aust. Can. Entomol. 115: 1359–1364.

Gutierrez, A. P., S. J. Yaninek, B. Wermelinger, H. R. Herren, and C. K. Ellis. 1988c. Analysis of the biological control of cassava pests in West Africa. III. The interaction of cassava and cassava green mite. J. Appl. Ecol. 25: 941–950.

Hairston, N. G., F. E. Smith, and L. B. Slobodkin. 1969. Community structure, population control, and competition. Am. Nat. 9: 421–425.

Hammond, W. M. O., P. Neuenschwander, and H. R. Herren. 1987. Impact of the exotic parasitoid (*Epidinocarsis lopezi*) on cassava mealybug (*Phenacoccus manihoti*) populations. Insect Sci. Appl. 8: 887–891.

Hardin, G. 1968. The tragedy of the commons. Science 162: 1243–1246.

Harper, J. L. and J. White. 1974. The demography of plants. Am. Rev. Ecol. Syst. 5: 419–463.

Hassell, M. P. 1978. *The Dynamics of Arthropod Predator–Prey Systems*. Princeton: Princeton University Press.

Hassell, M. P. 1986. Parasitoids and population regulation. In J. Waage and D. Greathead (eds.): *Insect Parasitoids* London: Academic Press, 389 pp.

Hassell, M. P., J. H. Lawton, and J. R. Beddington. 1976. The components of arthropod predation. I: The prey death-rate. J. Anim. Ecol. 45: 135–164.

Hassell, M. P., J. H. Lawton, and R. M. May. 1976. Patterns of dynamical behaviour in single species populations. J. Anim. Ecol. 45: 471–486.

Hassell, M. P. and R. M. May. 1974. Aggregation in predators and insect parasites and its effect on stability. J. Anim Ecol. 43: 567–594.

Hassell, M. P., R. M. May, S. W. Pacala, and P. L. Chesson. 1991. The persistence of host-parasitoid associations in patchy environments. I: A general criterion. Am. Nat. 138: 568–583.

Hassell, M. P. and G. C. Varley. 1969. New inductive population model for insect parasites and its bearing on biological control. Nature (London) 223: 1133–1137.

Hayman, B. I and A. D. Lowe, 1961. Transformation of counts of the cabbage aphid (Brevicoryne brassicae (L.)). N. Z. J. Sci. 4: 271–278.

Hegner, R. 1938. *Big Fleas Have Little Fleas, or Who's Who among the Protozoa.* Williams and Wilkins Co.

Hilborn, R. and D. Ludwig. 1993. The limits of applied ecological research. Ecol. Appl. 3: 350–352.

Hochberg, M. E., J. Pickering, and W. M. Getz. 1986. Evaluation of phenology models using field data: case study for the pea aphid, *Acyrthosiphon pisum*, and the blue alfalfa aphid, *Acrythosiphon kondoi* (Homoptera: Aphididae). Environ. Entomol. 15: 227–231.

Hoff, J. A. van't. 1884. *Etudes de Dynamique Chimique.* Amsterdam: Frederik Muller.

Holling, C. S. 1959. Some characteristics of simple types of predation and parasitism. Can. Entomol. 91: 293–320.

Holling, C. S. 1966. The functional response of invertebrate predators to prey density. Mem. Entomol. Soc. Can. 48: 3–86.

Holling, C. S. 1973. Resilience and Stability of Ecological systems. Ann. Rev. Ecol. Syst. 4: 1–23.

Holling, C. S. 1993. Investing in research for sustainability. Ecol. Appl. 3(4): 352–355.

Holling, C. S., D. Jones, and C. C. Clark. 1977. Ecological policy design: a case study of forest and pest management. In G. A. Norton and C. S. Holling (eds.): *Pest Management*, Oxford: International Institute for Applied Systems Analysis Proceedings Series, Pergamon, pp. 13–90.

Howard, L. O. and W. F. Fiske. 1911. The importation into the United States of the gypsy moth and the brown-tail moth. United States Dept. Agric. Bur. Entomol. Bull. 91, 312 pp.

Huffaker, C. B. 1944. The temperature relations of the immature stages of the malarial Mosquito, *Anopheles quadrimaculatus* Say, with a comparison of the developmental power of constant and variable temperatures in insect metabolism. Ann. Ent. Soc. Am. 37(1): 1–27.

Huffaker, C. B. 1958. Experimental studies on predation: dispersion factors and predator-prey oscillations. Hilgardia 27: 343–383.

Huffaker, C. B. 1980. *New Technology of Pest Control.* New York: Wiley, pp. 500.

Huffaker, C. B. and C. E. Kennett. 1959. A ten year study of the vegetational changes associated with biological control of Klamath weed. J. Range Man. 12: 69–82.

Huffaker, C. B., P. S. Messenger, and P. DeBach. 1971. The natural enemy component in natural control and the theory of biological control. In C. B. Huffaker (ed.): *Biological Control.* New York: Plenum Press, 477 pp.

Huffaker, C. B. and R. L. Rabb. 1984. *Ecology Entomology.* New York: Wiley, 844 pp.

Hughes, R. D. 1963. Population dynamics of the cabbage aphid *Brevicoryne brassicae* (L.). J. Anim. Ecol. 32: 393–426.

Hughes, R. D. and N. Gilbert. 1968. A model of an aphid population—a general statement. J. Anim. Ecol. 37: 533–563.

Hughes, R. D. and G. W. Maywald. 1990. Forecasting the favorableness of the Australian environment for the Russian wheat aphid, *Diuraphis noxia* (Homoptera: Aphididae), and its potential impact on Australian wheat yields. Bull. Entomol. Res. 80: 165–175.

Hutchinson, G. E. 1959. Homage to Santa Rosalia, or why there are so many kinds of animals? Am. Nat. 93: 145–159.

Huston, M., D. DeAngelis, and W. Post. 1988. New computer models unify ecological theory. Bioscience 38: 682–691.

Isley, D. 1932. Abundance of the boll weevil in relation to summer weather and food. Ark. Agric. Exp. Stan. Bull. 271.

Ivlev, V. S. 1955, *Experimental Ecology of the Feeding of Fishes.* (English translation by D. Scott 1961) New Haven: Yale University Press.

Janisch, E. 1925. Uber die Temperaturabhangigkeit Biologischer Vorgange und ihre Kurvenmabige Analyse. Pfluger Archiv fur die Gesamte Physiologie des Menschen und dir tiere. 209: 414–436.

Janisch, E. 1932. Trans. Entomol. Soc. London 80: 137–168.

Janssen, J. A. M. 1993. Soil nutrient availability in primary outbreak areas of the African armyworm, *Spodoptera exempta* (Lepidoptera: Noctuidae), in relationship to drought intensity and outbreak development in Kenya. Bull. Ent. Res. 83(9): 579–593.

Johnsen, S., A. P. Gutierrez, and J. Jorgensen (submitted) for publication. Overwintering in the cabbage root fly (*Delia radicum*: a dynamic model of temperature dependent dormancy and post dormancy development.

Jones, R. E., N. Gilbert, M. Guppy and V. Nealis. 1975. Long-distance movement of *Pieris Rapae.* J. Anim. Ecol. 49: 629–642.

Jones, R. E. 1977. Movement patterns and egg distribution in cabbage butterflies. J. Anim. Ecol. 46(1): 195–212.

Judson, O. P. 1994. The rise of the individual-based model in ecology. Trends in Ecology and Evol. 9: 9–14.

Karandinos, M. G. 1976. Optimal sample size and comments on some published formulae. Bull. Ent. Soc. Am. 22: 417–421.

Kelley, J. E. Jr. 1960. The Cutting Plane Method for Solving Convex Problems. Soc. Ind. Appl. Math. J. 8: 703–712.

Kettlewell, H. B. D. 1956. Further selection on industrial melanism in the Lepidoptera. Heredity London, 10: 287–301.

Kingsland, S. E. 1985. *Modeling Nature*. Chicago: University of Chicago Press.

Klay, A. 1987. Ecosysteme verger de pommier: Enquete faunistique sur les Phytoseiides et etude de leurs interactions avec l'acarien rouge *Panonychus ulmi* (Koch) en laboratoire. These EPFZ No. 8386, Zurich.

Krebs, C. J. 1978. *Ecology, the Experimental Analysis of Distribution and Abundance* (2nd edition), New York: Harper and Row.

Larson, E. B. and N. P. Thomsen. 1940. The Influence of Temperature on the Development of Some Species of Diptera. Series: Vidensk. Medd. Dan. Naturhist. Foren. Khobenharn. 104.

Law, J. 1983. A model for the dynamics of a plant population containing individuals classifed by age and size. Ecology 64, 224–230.

Lawton, J. 1977. Spokes missing in an ecological wheel. Nature (London) 265: 768.

Lawton, J. H., J. R. Beddington and R. Bonser. 1974. Switching in invertebrate predators. In M. B. Usher and M. H. Williamson (eds.): *Ecological Stability*. London: Chapman and Hall, pp. 141–158.

Lawton, J. H., M. P. Hassell and J. R. Beddington. 1975. Prey death rates and rate of increase of arthropod predator populations. Nature (London) 225: 60–62.

Lee, K. N. 1993. Greed, scale, mismatch, and learning. Ecol. Appl. 3: 360–364.

Leon, A. 1965. *A Comparison Among Eight Known Optimizing Procedures. Proceedings of the Symposium on Recent Advances in Optimizations Techniques*. In T. Vogl and A. Lavi (eds.): New York: Wiley.

Leibig, von J. 1840. *Chemistry and its Applications to Agriculture and Physiology*. London: Taylor and Walton (4th ed. 1847).

Leslie, P. H. 1945. On the use of matrices in certain population mathematics. Biometrika 33: 183–212.

Leslie, P. H. 1948. Some further notes on the use of matrices in population mathematics. Biometrika 35: 213–245.

Levins, R. 1975. Evolution of communities near equilibrium. In M. L. Cody and J. L. Diamond (eds.): *Ecology and Evolution of Communities*, Harvard University Press: Cambridge MA, pp. 16–50.

Levins, R. and M. Wilson. 1980. Ecological theory and pest management. Ann. Rev. Entomol. 25: 287–308.

Llewellyn, M. 1988. Aphid Energy Budgets. In *Aphids: Their Biology, Natural Enemies, and Control* (vol. B). Amsterdam: Elsevier, pp. 109–117.

Logan, J. A., R. E. Stinner, R. L. Rabb, and J. S. Bacheler. 1979. A descriptive model for predicting spring emergence of *Heliothis zea* populations in North Carolina. Environ. Entomol. 8: 141–146.

Logan, J. A., D. J. Wollkind, S. C. Hoyt, and L. K. Tanogoshi. 1976. An analytical model of temperature dependent rate phenomena in arthropods. Environ. Entomol. 5: 1133–1140.

Loomis, R. S. and W. A. Williams. 1963. Maximum crop productivity: an estimate. Crop Sci. 3, 67–72.

Lorenz, E. N. 1963. Deterministic non-periodic flow. J. Atm. Sci. 20: 448–464.

Lotka, A. J. 1925. *Elements of Physical Biology*. Baltimore: Williams and Witkins. (Reissued as Elements of mathematical biology by Dover, 1956).

Luck, R. F. 1971. An appraisal of two methods of analyzing insect life tables. Can. Entomol. 103: 1261–1271.

Ludwig, D. 1993. Environmental sustainability: magic, science and religion in natural resource management. Ecol. Appl. 3: 355–358.

Ludwig, D., R. Hilborn, and C. Walters 1993. Uncertainty, resource exploitation, conservation: lessons from history. Science 260: 17–36.

Ludwig, D., D. D. Jones, and C. S. Holling. 1978. Qualitative analysis of insect outbreak systems: the spruce budworm and forest. J. Anim. Ecol. 47: 315–332.

MacArthur, R. H. 1955. Fluctuations of animal populations, and a measure of community stability. Ecology 36: 533–536.

Mackay P. A., R. J. Lamb, and M. A. Hughes. 1989. Sexual fundatrix-like morphs in asexual populations of the pea aphid (Homoptera: Aphididae). Environ. Entomol. 18: 111–117.

Mangel, M., R. J. Hofman, E. A. Norse, and J. R. Twiss Jr. 1993. Sustainability and ecological research. Ecol. Appl. 3: 373–375.

Manetsch, T. J. 1976. Time-varying distributed delays and their use in aggregate models of large systems. IEEE Trans. Syst. Man Cybern. 6: 547–553.

Manly, B. F. J. 1989. *Stage Structured Populations. Sampling analysis and simulation*. London: Chapman and Hall, 187 pp.

Manly, B. F. J., P. Miller, and L. M. Cook. 1972. Analysis of a selective predation experiment. Am. Nat. 106(952): 719–736.

Marlaka, M. E., N. E. Stone, and S. B. Vinson. 1988. Host–parasitoid dynamics in a heterogenous environment. *Proc. SCS Multiconference on Artificial Intelligence and Simulation: The Diversity of Applications.* San Diego, CA, pp. 228–233.

Matson, P. A. and A. A. Berryman. 1992. Special Features: ratio-dependent predator-prey theory. Ecology 73: 1531–1566.

Matson, P. A. and S. L. Ustin. 1991. Special Features: The future of remote sensing in ecological studies. Ecology 72: 1917.

May, R. M. 1973. *Stability and Complexity in Model Ecosystems*. Princeton University Press, Princeton, NJ.

May, R. M. 1978. Host–parasitoid systems in patchy environments: a phenomenological model. J. Anim. Ecol. 47: 833–843.

May, R. M. (ed.) 1981. *Theoretical Ecology: principles and application.* (second edition). Sunderland, MA: Sinauer Associates, 489 pp.

May, R. M. and G. Oster. 1976. Bifurcations and dynamics complexity in simple ecological models. Am. Nat. 100: 573–599.

Mayer, A. and J. Mayer. 1974. Agriculture: the islands' empire. In *Science and Its Public; the Changing Relationship*. Daedalus vol. 103 (3) pp. 83–95.

Maynard Smith, J. 1974. *Models in Ecology.* Cambridge: Cambridge University Press.

McIntosh, R. P. 1987. Pluralism in Ecology. Ann. Rev. Ecol. Syst. 18: 321–341.

McKendrick, A. G. 1926. The application of mathematics to medical problems. Proc. Edinburgh Math. Soc. 44: 98–130.

McNeill, S. and T. R. E. Southwood. 1978. The role of nitrogen in the development of insect/plant relationships. In J. B. Harborne (ed.): *Biochemical Aspects of Plant and Animal Coevolution*, London: Academic Press, pp. 77–98.

Messenger, P. S. 1964. Use of life-tables in a bioclimatic study of an experimental aphid-braconid wasp host–parasite system. Ecology 45: 119–131.

Messenger, P. S. 1968. Bioclimatic studies of the aphid parasite *Praon exsoletum.* 1. effects of temperature on the functional response of females to varying host densities. Can. Entomol. 100: 728–741.

Messenger, P. S. and D. C. Force. 1963. An experimental host-parasite system: *Therioaphis maculata* (Buckton), *Praon palitans* Musebeck (Homoptera: Aphididae-Hymenoptera: Bracondidae). Ecology 44: 532–540.

Metzgar, L. H. and E. Boyd. 1988. Stability properties in a model of forage-ungulate-predator interactions. Nat. Res. Modeling 3: 3–43.

Meyer, J. L. and G. S. Helfman. 1993. The ecological basis of sustainability. Ecol. Appl. 3: 369–371.

Meyer, J. S., C. G. Ingersoll, L. L. McDonald, and M. S. Boyce. 1986. Estimating uncertainty in population growth rates: jackknife vs. bootstrap techniques. Ecology 67: 1156–1166.

Mills, N. J. 1981. Some aspects of the rate of increase of a coccinellid. Ecol. Entomol. 6: 293–299.

Mills, N. J. and A. P. Gutierrez. 1996. Prospective modeling in biological control: an analysis of the dynamics of heteronomous hyperparasitism. J. Anim. Ecol.

Milne, A. 1959. A controversial equation in population ecology. Nature (London) 184: 1582–1583.

Monteith, J. L. 1965. Light distribution and photosynthesis in field crops. Ann. Botany NSW 29: 17–37.

Mooney, H. A. and S. E. Sala. 1993. Science and sustainable use. Ecol. Appl. 3(4): 564–566.

Morris, R. F. 1963. Predictive population equations based on key factors. Mem. Entomol. Soc. Can. 32: 16–21.

Murdoch, W. W. 1969. Switching in general predators: experiments on predator specificity and stability of prey populations. Ecol. Mon. 39: 335–354.

Murdoch, W. W. 1991. The shift from equilibrium to a non-equilibrium paradigm in ecology. Bull. Ecol. Soc. Am. 72: 49–51.

Murdoch, W. W. 1994. Population regulation in theory and practice. Ecology 75: 271–287.

Murdoch, W. W., J. Chesson, and P. L. Chesson. 1985. Biological control in theory and practice. Am. Nat. 125: 344–366.

Murdoch, W. W. and J. D. Reeve. 1987. Aggregation of parasitoids and the detection of density dependence in field populations. OIKOS 49: 1–5.

Munster-Swendsen, M. 1985. A simulation study of primary-, clepto- and hyper-parasitism in *Epinotia tedella* (Lepidoptera, Tortricidae). J. Anim. Ecol. 54: 683–695.

Munster-Swendsen, M. 1991. The effect of sublethal neogregarine infections in the spruce needleminer, *Epinotia tedella* (Cl.) (Lepidoptera: Torticidae). Ecol. Entom. 16: 211–219.

Nachman, G. 1982. A mathematical model of the functional relationship between density and spatial distribution of a population. J. Anim. Ecol. 50: 453–460.

Nachman, G. 1984. Estimates of mean population density and spatial distribution of *Tetranychus urticae* (Acarina: Tetranychidae) and *Phytoseiulus persimillis* (Acarina, Phytoseiidae) based upon the proportion of empty sampling units. J. Appl. Ecol. 21: 903–913.

Nelder, J. A. and R. Mead. 1965. A Simplex method for function minimization. Comp. J. 7: 308–313.

Neuenschwander, P., W. N. O. Hammond, A. P. Gutierrez, A. R. Cudjoe, J. U. Baumgärtner, U. Regev, and R. Adjakloe. 1989. Impact assessment of the biological control of the cassava mealybug, *Phenacoccus manihoti* Matile Ferrero (Hemiptera: Pseudococcidae) by the introduced parasitoid, *Epidinocarsis lopezi* (DeSantis) (Hymenoptera: Encyrtidae). Bull. Ent. Res. 79: 579–594.

Neuenschwander, P., W. N. O. Hammond, and R. D. Hennessey. 1987. Changes in the composition of the fauna associated with the cassava mealybug, *Phenacoccus manihoti*, following the introduction of the parasitoid *Epidinocarsis lopezi* (DeSantis). In Africa-wide Biological Control Project of Cassava Pests. In P. Neuenschwander, J. S. Yaninek and H. R. Herren (eds.): Insect Scientific Application (special issue) 8: 893–898.

Neuenschwander, P. and E. Madojemu. 1986. Mortality of the cassava mealybug, *Phenacoccus manihoti* Mat.-Ferr. (Hom., Pseudococcidae), associated with an attack by *Epidinocarsis lopezi* (Hym., Encyrtidae). Mitt. Schweiz. Ent. Ges. 59: 57–62.

Neuenschwander, P., F. Schulthess, and E. Madojemu. 1986. Experimental evaluation of the efficiency of *Epidinocarsis lopezi*, a parasitoid introduced into Africa against the cassava mealybug *Phenacoccus manihoti*. Entomol. Exp. App. 42: 27–32.

Nicholson, A. J. 1933. The balance of animal populations. J. Anim. Ecol. 2: 131–178.

Nicholson, A. J. 1954a. Compensatory relation of animal populations to stresses and their evolutionary significance. Aust. J. Zoo. 2: 1–8.

Nicholson, A. J. 1954b. An outline of the dynamics of animal populations. J. Aust. Zoo. 2: 9–65.

Nicholson, A. J. 1957. The self adjustment of populations to change. Cold Springs Harbor Symp. on Quant. Biol. 22: 153–597.

Nicholson, A. J. and V. A. Bailey. 1935. The balance of animal populations. Part 1. Proc. Zoo. Soc. London 3: 551–598.

Nicole, S. and W. de la Mare. 1993. Ecosystem management and the antarctic krill. Am. Sci. 81: 36–47.

Nowierski, R. M., A. P. Gutierrez, and J. S. Yaninek. 1983. Estimation of thermal thresholds and age-specific life table parameters for the walnut aphid (Homoptera: Aphididae) under field conditions. Environ. Entomol. 12: 680–686.

Odebiyi, J. A. and A. H. Bokonon-Ganta. 1986. Biology of *Epidinocarsis [= Apoanagyrus] lopezi* [Hymenoptera: Encyrtidae] an exotic parasite of cassava mealybug, *Phenacoccus manihoti* [Homoptera: Pseudococcidae] in Nigeria. Entomophaga 31: 251–260.

Odum, E. P. 1971. *Fundamentals of Ecology*, 3rd ed. Toronto: W. B. Saunders and Co.

Olson, R. L., P. J. H. Sharpe, and H. Wu. 1985. Whole plant modelling: a continuous-time Markov (CTM) approach. Ecol. Modelling 29: 171–187.

O'Neil, R. J. 1988. A Model of Predation by *Podisus maculventris* (Say) on Mexican Bean Beetle, *Epilachna varivestis* Mulsant, in Soybeans. Can. Entomol. 120: 601–608.

O'Neil, R. J. 1989. Comparison of laboratory and field measurements of the functional response of *Podisus maculiventris* (Heteroptera: Pentatomidae). J. Kansas Ent. Soc. 62: 148–155.

Oster, G. 1981. Predicting populations. Am. Zool. 21:831–844.

Oster, G. and A. Ipaktchi. 1978. In Eyring, H. and D. Henderson (eds.): Theoretical Chemistry: Periodicity in Chemistry and Biology, New York: Academic Press, pp. 111–132.

Pearl, R. and L. J. Reed. 1920. On the rate of growth of the population of the United States since 1790 and its mathematical representation. Proc. Natl. Acad. Sci., U.S.A., 6: 275–288.

Penman, H. L. 1948. Natural Evaporation from open water, bare soil and grass. Proc. R. Soc. Ser. A. 193: 120–145.

Peterman, R. M., W. C. Clark, and C. S. Holling. 1979. The dynamics of resilience: shifting stability domains in fish and insect systems. In R. M. Anderson, B. D. Turner, and L. R. Taylor (eds.): *Population Dynamics*. London: Blackwell Scientific Publications.

Peters, R. H. 1983. *The ecological implications of body size*. Cambridge: Cambridge University Press, 329 pp.

Petrusewicz, K. and A. MacFayden, 1970. *Productivity of Terrestrial Animals: Principles and Methods*. IBP Handbook 13. Oxford: Blackwell.

Phillipson, J. 1966. *Ecological Energetics*. London: Edward Arnold.

Pianka, E. 1981. Resource acquisition and allocation among animals. In C. R. Townsend and P. Calow (eds.): *Physiological Ecology: an evolutionary approach to resource use*. Sinauer Associates, Sunderland, MA, pp. 300–314.

Pickett, C. H., L. T. Wilson, D. L. Flaherty, and D. Gonzalez. 1987. Biological control of variegated leafhopper. Calif. Agric. 41: 14–16.

Pickett, C. H., L. T. Wilson, D. L. Flaherty, and D. Gonzalez. 1989. Measuring the host preference of parasites: an aid in evaluating biotypes of *Anagrus epos* [Hym.: Mymaridae]. Entomophaga 34: 551–558.

Pielou, E. C. 1969. *An introduction to mathematical ecology*. New York: Wiley-Interscience, 286 pp.

Pijld, J. W. A. M., K. D. Kofker, M. J. van Staalduinen, and J. J. M. van Alphen. 1990. Interspecific host discrimination and competition by *Epidinocarsis lopezi* and *E. diversicornis*, parasitoids of the cassava mealybug, *Phenacoccus manihoti*. Medical Faculty Landbouww. Rijksuniversity. Gent. 55: 404–416.

Pimm, S. L. 1982. *Food Webs*. London: Chapman and Hall.

Pitelka, L. F. and F. A. Pitelka. 1993. Environmental decision making: multidimensional dilemmas. Ecol. Appl. 3: 366–368.

Pizzamiglio, M. A., A. P. Gutierrez, W. J. dos Santos, K. D. Gallagher, W. S. Oliveira, and Z. V. Fujita. 1989. Phenological patterns and sampling decision rules for arthropods in cotton fields in Parana, Brazil: before boll weevil. Pesqui. Agropecu. Bras. 24: 337–345.

Plant, R. E. and L. T. Wilson. 1986. Models for age structured populations with distributed maturation rates. J. Math. Biol. 23: 247–262.

Polak, E. 1973. An historical survey of computational methods in optimal control. Soc. Indust. Appl. Math. (SIAM) Rev. 15: 553–584.

Policanski, D. 1993. Uncertainty, knowledge and resource management. Ecol. Appl. 3: 383–384.

Powell, M. J. D. 1964. An efficient method for finding the minimum of a function of several variables without calculating the derivatives. Comp. J. 7: 155–162.

Powell, M. J. D. 1970. A survey of numerical methods for unconstrained optimization. SIAM Rev. 12: 79–97.

Price, P. W. 1984. *Insect Ecology* (second edition). New York: Wiley.

Randolph, P. A., J. C. Randolph and C. A. Barlow. 1975. Age specific energetics of the pea aphid—*Acyrthosiphon pisum*. Ecology 56: 359–369.

Readshaw, J. L. 1981. The glass bead game. Nature (London): 292: 178.

Readshaw, J. L. and W. R. Cuff. 1980. A model of Nicholson's blowfly cycles and its relevance to predation theory. J. Anim. Ecol. 49: 1005–1010.

Real, L. A. 1979. Ecological determinants of functional response. Ecology 60: 481–485.

Reeves, J. D. and W. W. Murdoch. 1985. Aggregation by parasitoids in the successful control of the red scale insect: a test of theory. J. Anim. Ecol. 54:797–816.

Reeves, J. D. and W. W. Murdoch. 1986. Biological control by the parasitoid *Aphytis melinus*, and population stability of the California red scale. J. Anim. Ecol. 55: 1069–1082.

Regev, U., A. P. Gutierrez, and G. Feder. 1976. Pests as a common property resource: a case study in the control of the alfala weevil. J. Agric. Econ. 58: 186–97.

Regev, U., A. P. Gutierrez, and J. E. DeVay. 1990. Optimal strategies for management of verticilium wilt. Agric. Syst. 33: 139–152.

Richtmyer, R. D. and K. W. Morton. 1967. *Finite difference methods for initial value problems*. New York, London and Sydney: Interscience Publishers.

Ritchie, J. T. 1972. Models for predicting evaporation from a crop with incomplete cover. Water Resources Research 8: 1204–1213.

Rogers, D. 1972. Random search and insect population models. J. Anim. Ecol. 41: 369–383.

Roffey J. and G. Popov. 1968. Environmental and behavioural processes in Desert locust outbreaks. Nature (London) 219: 446–450.

Rosen, J. B. 1960. The Gradient projection method for nonlinear programming. Part I. Linear Constraints. SIAM J. 8: 181–217.

Rosen, J. B. 1961. The gradient projection method for nonlinear programming. Part II. Nonlinear Constraints. SIAM J. 9: 514–532.

Rosenheim, J. A., T. Meade, I. G. Powch, and S. E. Schoenig. 1989. Aggregation by foraging insect parasitoids in response to local variations in host density: determining the dimensions of a host patch. J. Anim. Ecol. 58: 101–117.

Rosenzweig, M. L. and R. H. MacArthur. 1963. Graphical representation and stability conditions of predator–prey interactions. Am. Nat. 97: 209–23.

Roughgarden, J. 1979. Theory of Population Genetics and Evolutionary Ecology: an Introduction. New York: MacMillan, 634 pp.

Roughgarden, J., S. W. Running, and P. A. Matson. 1991. What does remote sensing do for ecology. Ecology 72(6): 1918–1922.

Royama, T. 1971. A comparative study of models for predation and parasitism. Res. Popul. Ecol. Kyoto Univ. Suppl. 1: 1–91.

Royama, T. 1981. Evaluation of mortality factors in insect life-tables analysis. Ecol. Mono. 543: 429–62.

Royama, T. 1992. *Analytical Population Dynamics.* London: Chapman and Hall, 371 pp.

Rubenstein, D. I. 1993. Science and the pursuit of a sustainable world. Ecol. Appl. 3: 385–387.

Ruesink, W. G. and M. Kogan. 1982. The quantitative basis of pest management: sampling and measuring, In R. L. Metcalf and W. H. Luckman, (eds.): *Introduction to Insect Pest management*, New York: Wiley, pp. 315–352.

Ruxton, G. D. and W. S. Gurney. 1992. The interpretation of a test for ratio-dependence. Oikos 65: 334–335.

Salwasser, H. 1993. Sustainability needs more than better science. Ecol. Appl. 3: 387–389.

Schaffer, W. M. 1985. Order and Chaos in ecological systems. Ecology 66: 93–106.

Schaffer, W. M. and M. Knot. 1986. Chaos in ecological systems: the coals that Newcastle forgot. Trends in Ecology and Evol. 1: 58–63.

Schoolfield, R. M., P. J. H. Sharpe, and C. E. Magnuson. 1981. Nonlinear regression of biological temperature dependent rate models based on absolute reaction rate theory. J. Theor. Biol. 88: 719–731.

Schreiber, S. R. and A. R. Gutierrez. 1996. A supply/demand perspective of persistence in food webs: applications to biological control. Ecology (submitted).

Scriber, J. M. and P. Feeny. 1979. Growth of herbivorous caterpillars in relation to feeding specialization and to the growth form of their food plants. Ecology 60: 829–850.

Sebens, K. P. 1987. The Ecology of indeterminate growth in animals. Ann. Rev. Ecol. Syst. 18: 371–407.

Sequeira, R. A., P. J. H. Sharpe, N. D. Stone and M. E. Markela. 1991. Object oriented simulation: plant growth and discrete organ to organ interactions. Ecol. Modelling 58: 55–89.

Severini, M., J. Baumgärtner, and M. Ricci. 1990. Theory and practice of parameter estimation of distributed delay models for insect and plant phenologies. In R. Guzzi, R. A. Navarra, and J. Shukla (eds.): *Meteorology and Environmental Sciences.* Singapore: World Scientific and International Publisher, pp. 674–719.

Sharpe, P. J. H., G. L. Curry, D. W. DeMichele, and C. L. Cole. 1977. Distribution model of organism developmental times. J. Theor. Biol. 66: 21–38.

Sharpe, P. J. H. and D. W. DeMichele, 1977. Reaction kinetics of poikilotherm development. J. Theor. Biol. 64: 649–670.

Sharpe, P. J. H. and L. C. Hu. 1980. Reaction kinetics of nutrition dependent poikilotherm development. J. Theor. Biol. 82: 317–333.

Shelford, V. E. 1931. Some concepts of bioecology. Ecology 12: 455–467.

Sinko, J. W. and W. Streifer. 1967. A new model for age-structure of a population. Ecology 48: 910–918.

Slobodkin, L. B. 1961. *Growth and regulation of Animal Populations.* New York: Holt, Rinehart, and Winston, 184 pp.

Slobodkin, L. B. 1993. Scientific goals require literal empirical assumptions. Ecol. Appl. 3: 371–373.

Smale, S. 1976. A convergent process of price adjustment and global newton methods. J. Math. Econ. 3: 107–120.

Smith, H. S. 1935. The role of biotic factors in the determination of population densities. J. Econ. Ent. 28: 873–898.

Smith, H. S. 1939. Insect populations in relation to biological control. Ecol. Monog. 9: 311–320.

Solomon, M. E. 1949. The natural control of animal populations. J. Anim. Ecol. 18: 1–35.

Socolow, R. H. 1993. Achieving sustainable development that is mindful of human imperfections. Ecol. Appl. 3: 381–383.

Southwood, T. R. E. 1975. *Ecological Methods.* London: Butler and Tanner, 383 pp.

Southwood, T. R. E., M. P. Hassell, P. M. Reader, and D. J. Rogers. 1989. Population dynamics of the viburnum whitefly (*Aleurotrachelus jelinekii*). J. Anim. Ecol. 58: 921–942.

Spendly, W., G. R. Hext and F. R. Himsworth. 1962. Sequential application of simplex designs in optimisation and evolutionary operation. Technometrics: 4: 441–461.

Stevens, R. T. 1990. Fractals: Programming in TURBO PASCAL. Redwood City, CA: M. and T. Publ., 462 pp.

Stewart, G. W. III. 1967. A modification of Davidon's minimization method to accept difference approximation of derivatives. J. Assoc. Comput. Mach. 14: 72–83.

Stinner, R. E., G. D. Butler, J. S. Bacheler, and C. Tuttle. 1975. Simulation of temperature dependent development in population dynamics models. Environ. Entomol. 107: 1167–1174.

Stinner, R. E., A. P. Gutierrez and G. D. Butler. 1974. An algorithm for temperature-dependent growth rate simulation. Can. Ent. 106: 519–524.

Stone, N. D. and A. P. Gutierrez. 1986. Pink Bollworm control in southwestern desert cotton. I. A field-oriented simulation model. Hilgardia 54: 1–24.

Streifer, W. 1974. Realistic models in population ecology. *Advances in Theoretical Ecology.* 8: 199–266.

Streifer, W. and C. A. Istock, 1973. A Critical variable formulation of population dynamics. Ecology 54: 393–398.

Strong, D. R., Jr. 1983. Natural variability and the manifold mechanisms of ecological communities. Am. Nat. 122: 636–660.

Sugihara G. and R. M. May. 1990. Nonlinear forecasting as a way of distinguishing chaos from measurement error in time series. Nature (London) 344: 735–741.

Sutherst, R. W., G. F. Maywald, and W. Bottomly. 1991. From CLIMEX to PESKY, a generic expert system for risk assessment. EPPO Bulletin 21: 595–608.

Summers C. G., R. L. Coviello, and A. P. Gutierrez. 1984. Influence of constant temperatures on the development and reproduction of *Acyrthosiphon kondoi* (Homoptera: Aphididae) Environ. Entomol. 13: 236–242.

Symmons, P. 1992. Strategies to combat the desert locust. Crop Prot. 11: 206–212.

Tamò, M. and J. Baumgärtner. 1993. Analysis of the cowpea ecosystem in West Africa. I. A demographic model for carbon acquisition and allocation in cowpea, *Vigna unguiculata* (L.) Walp. Ecol. Modelling 65: 95–121.

Tauber, M. J., C. A. Tauber, and S. Masaki. 1986. Seasonal adaptations in Insects. Oxford University Press, 411 pp.

Taylor, C. E. and R. R. Sokal. 1976. Oscillations in housefly population sizes due to time lags [*Musca domestica*]. Ecology 57: 1060–1067.

Taylor, F. 1981. Ecology and evolution of physiological time in insects. Am. Nat. 117: 1–23.

Taylor, L. R. 1961. Aggregation, variance, and the mean. Nature (London) 189: 732–735.

Thompson, W. R. 1939. Biological control and the theories of the interaction of populations. Parasitology 31: 299–388.

Tinbergen, N. 1960. *The Study of Instinct*. Oxford: Clarendon Press.

Townsend, C. R. and P. Calow. 1981. *Physiological Ecology: An evolutionary approach to resource use*. Sunderland, MA: Sinauer Associates, 393 pp.

Tucker, C. J., R. G. Townshend, and T. E. Goff. 1985. African land cover classification using satellite data. Science 227: 369–375.

Tyndale-Briscoe M. and R. D. Hughes. 1969. Changes in the female reproductive system as age indicators in the bushfly *Musca vetustissima* Wlk. Bull. Entomol. Res. 59: 129–141.

U.S. Congress, Office of Technology Assessment. 1990. A Plague of Locusts-Special Report, OTA-F-450 (Washington, DC: US Government Printing Office).

van den Bosch, R. 1978. *The Pesticide Conspiracy*. New York: Doubleday.

van den Bosch, R., B. D. Frazer, C. S. Davis, P. S. Messenger, and R. Hom. 1970. An effective walnut aphid parasite from Iran. Calif. Agric. 24: 8–10.

van den Bosch, R., P. S. Messenger and A. P. Gutierrez. 1982. Introduction to Biological Control. New York: Plenum Press.

van den Bosch, R., E. I. Schlinger, E. J. Dietrick, J. C. Hall, and B. Puttler. 1964. Studies on the succession, distribution and phenology of imported parasites of *Therioaphis trifolii* (Monell) in Southern California. Ecology 45: 602–621.

van den Meiracker, R. A. F., W. N. O. Hammond, and J. J. M. van Alphen. 1988. The role of kairomones in prey finding in the coccinellids *Diomus sp.* and *Exochomus sp.*, predators of the cassava mealybug, *Phenacoccus manihoti*. Med. Fac Landouww. 53(3a): 1063–1077.

van der Waerden, B. L. 1953. *Modern Algebra*. New York: Fredrick Ungar.

Vansickle, J. 1977. Attrition in distributed delay models. IEEE Trans. Sys., Man, Cybern. 7: 635–638.

Varley, C. G. and G. R. Gradwell. 1960. Key factors in population studies. J. Anim. Ecol. 29: 399–401.

Varley, C. G., G. R. Gradwell, and M. P. Hassell. 1973. *Insect Population Ecology: an analytical approach.* Berkeley: University of California Press, 212 pp.

Villacorta, A. M. and A. P. Gutierrez. 1989. Presence-absence decision rules for the damage caused by the coffee-leaf miner (*Leucoptera coffeela* (Guerin-Meneville 1842)). Pesqui. Agropecu. Bras. 24: 517–525.

Verhulst, P. E. 1938. Notice sur la loi que la population suit dans acroissement. Corresp. Math. Phys. 10: 113–126.

Vinson, S. B. 1981. Habitat location. In (D. A. Nordlund, R. L. Jones and W. A. Lewis, eds.): *Semiochemicals: their role in pest control.* New York: Wiley, pp. 51–78.

Volterra, V. 1926. Variations and fluctuations of the number of individuals in animal species living together. J. Cons. perm. int. Ent. Mer. 3: 3–51. (reprinted in Chapman, R. N. 1931. *Animal Ecology*, New York and London.)

Volterra, V. 1931. Principles de biologie mathematique. Acta Biotheor. 3: 1–36.

Walters, C. 1986. Adaptive Management of Renewable Resources. New York: Macmillan.

Wang, Y. H. and A. P. Gutierrez. 1980. An assessment of the use of stability analyses in population ecology. J. Anim. Ecol. 49: 435–452.

Wang, Y., A. P. Gutierrez, G. Oster, and R. Daxl. 1977. A population model for cotton growth and development: coupling cotton-herbivore interaction. Can. Entomol. 109: 1359–1374.

Ward, S. A., A. F. G. Dixon and P. W. Welling. 1983b. The relation between fecundity and reproductive investment in aphids. J. Anim. Ecol. 52: 451–461.

Ward, S. A., P. W. Wellings, and A. F. G. Dixon. 1983a. The effect of reproductive investment on pre-reproductive mortality in aphids. J. Anim. Ecol. 52: 305–313.

Watt, K. E. F. 1959. A mathematical model for the effects of densities of attacked and attacking species on the number attacked. Can. Entomol. 91: 129–144.

Weigert, R. G. 1976. *Benchmark Papers in Ecology: Ecological Energetics.* Stroudsberg, PA: Dowden, Hutchingson and Ross.

Welch, S. M., B. A. Croft, J. F. Brunner, and M. F. Michels 1978. PETE: an extension phenology modeling system for management of a multi-species complex. Environ. Entomol. 7, 487–494.

Wermelinger, B., J. Baumgärtner and A. P. Gutierrez. 1991a. A demographic model for assimilation and allocation of carbon and nitrogen in grapevines. Ecol. Modelling. 53: 1–26.

Wermelinger, B., J. J. Oertli, and J. Baumgärtner. 1991b. Environmental factors affecting the life table statistics of *Tetranychus urticae* (Acari: Tetranychidae). III. Host plant nutrition. Exp. Appl. Acarology. 12: 259–274.

Weseloh, R. M. 1981. Host location by parasitoids. In D. A. Nordlund, R. L. Jones, and W. A. Lewis (eds.): *Semiochemicals: their role in pest control.* New York: Wiley, pp. 79–95.

Westphal, D. F., A. P. Gutierrez, and G. D. Butler Jr. 1979. Some interactions of the pink bollworm and cotton fruiting structures Hilgardia 47: 177–190.

White, E. B., P. De Bach, and M. J. Garber. 1970. Artificial selection to temperature extremes in *Aphytis lingnaesis* Compere (Hymenoptera: Aphelinidae). Hilgardia 40: 161–192.

Wickland, D. E. 1991. Mission to planet earth: the ecological perspective. Ecology 72: 1923–1933.

Williams, D. W. 1984. Ecology of the blackberry-leafhopper-parasite system and its relevance to California grape ecosystems. Hilgardia 52: 1–32.

Wilson, L. T. and A. P. Gutierrez, 1980. Within plant distribution of predators on cotton: comments on sampling and predator efficiencies. Hilgardia 48: 3–11.

Wilson, L. T., C. Pickel, R. C. Mount, and F. G. Zalom. 1983. Presence-absence sequential sampling for cabbage aphid and green peach aphid (Homoptera: Aphididae) on Brussels sprouts. J. Econ. Entomol. 76: 476–479.

Wilson, L. T. and P. M. Room. 1982. The relative efficiency and reliability of three methods for sampling arthropods in Australian cotton fields. J. Austr. Entomol. Soc. 21: 175–181.

Wilson, L. T., J. M. Smilanick, P. Hoffman, D. L. Flaherty, and S. M. Ruiz. 1988. Leaf nitrogen and position in relation to population parameters of Pacific spider mite, *Tetranychus pacificus* (Acari: Tetranychidae) on grape. Environ. Entomol. 17: 964–968.

Wilson, L. T., W. L. Sterling, D. R. Rummel, and J. E. De Vay. 1989. Quantitative sampling principles in cotton IPM. In R. F. Frisbie, K. M. El-Zik, and L. T. Wilson. *Integrated Pest Management and Cotton Production.* New York: Wiley, pp. 85–119.

Winters, S. F. 1971. Satisficing, selection, and the innovating remnant. Quart. J. Econ. 85: 237–261.

de Wit, C. T. and J. Goudriaan 1978. *Simulation of Ecological Processes* 2nd ed. The Netherlands: PUDOC Publishers.

Yaninek, J. S. 1985. An assessment of the phenology, dynamics and impact of the cassava green mite on cassava yields in Nigeria: a component of biological control. Thesis (Ph.D.), University of California, Berkeley.

Yaninek, J. S., A. P. Gutierrez and H. R. Herren. 1989. Dynamics of *Mononychellus tanajoa* (Acari: Tetranychidae) in Africa: experimental evidence of temperature and host plant effects on population growth rates. Environ. Entomol. 7: 499–501.

Zahner, P. and J. Baumgärtner. 1988. Analyse des Interactions Plante-Tetranyques-Phytoseides. I. Modèles de Population la Dynamique de *Panonychus ulmi* (Koch) et *Tetranychus urticae* Koch en Vergers de Pommiers. Acta Oecolog, Oecol. Applic. 9: 311–331.

Zedler, J. B. 1993. Lessons on preventing overexploitation. Ecol. Appl. 3: 377–378.

Zoutendijk, G. 1960. *Methods of Feasible Directions: A Study in Linear and Nonlinear Programming.* Amsterdam: Elsevier.

Author Index

289

Subject Index